World Sustainability Series

Due to its scope and nature, sustainable development is a matter which is very interdisciplinary, and draws from knowledge and inputs from the social sciences and environmental sciences on the one hand, but also from physical sciences and arts on the other. As such, there is a perceived need to foster integrative approaches, whereby the combination of inputs from various fields may contribute to a better understanding of what sustainability is, and means to people. But despite the need for and the relevance of integrative approaches towards sustainable development, there is a paucity of literature which address matters related to sustainability in an integrated way.

Notes on the quality assurance and peer review of this publication

Prior to publication, the works published in this book are initially assessed and reviewed by an in-house editor. If suitable for publication, manuscripts are sent for further review, which includes a combined effort by the editorial board and appointed subject experts, who provide independent peer-review. The feedback obtained in this way was communicated to authors, and with manuscripts checked upon return before finally accepted. The peer-reviewed nature of the books in the "World Sustainability Series" means that contributions to them have, over many years, been officially accepted for tenure and promotion purposes.

Christian Brandt · Esteban Mejía · Bettina Schorr
Editors

Implementing Sustainability and Internationalization in Higher Education

The Case of the SDG Graduate Schools

Springer

Editors
Christian Brandt
Institute of Farm Management
University of Hohenheim
Stuttgart, Germany

Esteban Mejía
Leibniz Institute for Catalysis (LIKAT)
Rostock, Germany

Bettina Schorr
Institute for Latin American Studies
Freie Universität Berlin
Berlin, Germany

ISSN 2199-7373 ISSN 2199-7381 (electronic)
World Sustainability Series
ISBN 978-3-031-97836-4 ISBN 978-3-031-97837-1 (eBook)
https://doi.org/10.1007/978-3-031-97837-1

This Springer imprint is published by the registered company Springer Nature Switzerland AG
The registered company address is: Gewerbestrasse 11, 6330 Cham, Switzerland

This book is dedicated to the loving memory of ***Avhatakali (Taki) Sithagu****, whose spirit, wisdom, and warmth continue to inspire all who knew her. May her legacy live on in the pursuit of knowledge, compassion, and a more sustainable world.*

Foreword

The 2030 Agenda for Sustainable Development presents an ambitious vision for a more just, equitable, and sustainable world. However, the urgency of the sustainability challenges we face, including climate change, biodiversity loss, and social inequality, demands accelerated action. Achieving the Sustainable Development Goals (SDGs) requires concerted efforts from all sectors of society. Higher education—with its core missions of research, innovation, teaching, and civic engagement—is a crucial component of this endeavour.

DAAD has actively been promoting higher education partnerships for the SDGs through the programme "SDG Graduate Schools" since 2016. This programme, funded by the German Federal Ministry for Economic Cooperation and Development (BMZ), supports the establishment of seven Graduate Schools in low- and middle-income countries. By fostering collaborations between universities in Africa, Asia, and Latin America, it aims to strengthen higher education systems and build capacity for sustainable development. The needs-based, bottom-up approach has empowered partner universities with significant flexibility to tailor their projects to local contexts. The funding covers a broad spectrum, ranging from scholarships and short-term stays to activities aimed at establishing sustainable university structures. Experts and lecturers at master's, doctoral and post-doctoral level are thus enabled to work on innovative solutions for key SDG topics in line with demand.

Through their research, teaching, co-creation and outreach activities, the participating higher education institutions have made significant contributions to overcoming complex global challenges. Their partnerships have led to increased research capacity, improved curriculum, and highly qualified professionals in development related fields. Even during times of political crises, pandemics, and other adversities, the SDG Graduate Schools have shown remarkable resilience. The integrated approach of individual and institutional support has proven to be especially effective in fostering transformation and delivering sustainable outcomes.

External evaluations have shown that the SDG Graduate Schools are highly relevant for national and international development agendas and have achieved remarkable results. Since the programme was launched in 2016, over 200 courses and study programmes have been revised or newly developed. Significant research outputs provide evidence of the Graduate Schools' successful international collaboration and joint knowledge creation. More than 1200 publications have been issued, with about two thirds of them appearing in peer-reviewed journals. Since the SDG Graduate Schools started, nearly 800 scholarships have been awarded, most of them on Master and PhD-Levels. In addition, almost 10,000 participants benefited from further education and training events offered by the SDG Graduate Schools. Over the years, a large proportion of former scholarship holders have built successful careers in development.

Global challenges such as climate change and food security are closely interwoven. That is why the 2030 Agenda attaches great importance to ensuring that social, economic, and environmental dimensions of sustainability are considered together, not weighed against each other. In line with this need for an integrated approach, the SDG Graduate Schools have been at the forefront of tackling complex interlinkages between the SDGs. By bringing together multiple perspectives and expertise from around the world, they have explored the synergies and trade-offs between different goals. This perspective allows for a more holistic understanding of sustainable development and enables the identification of more effective solutions.

Interdisciplinary and transdisciplinary approaches are key to achieving the SDGs. The German Advisory Council on Global Change (WGBU) describes the necessary shift towards transformative science as a "societal search process" that requires both the initiative and the will to change on the part of all stakeholders involved. It is not only researchers who have to cross methodological and geographical boundaries; universities, funding organisations, and ministries are also challenged to reposition themselves in this regard. The SDG Graduate Schools have actively embraced the growing need for transnational collaboration, employing interdisciplinary and transdisciplinary approaches to address societal challenges. Despite the diverse trajectories of specialized disciplines, they interact in complex ways within the context of the sustainability agenda, exploring new forms of knowledge production and working with external partners and stakeholders on concrete, problem-oriented solutions.

In this respect, the establishment of the "SDG Graduate Schools Alliance" in 2020 marks an important milestone: It has since facilitated greater networking and collaboration among the projects, enabling them to harness synergies, foster interdisciplinary capacity building, and drive forward knowledge transfer in both substantive and methodological terms. The joint Midterm Conference—Digitalization in Higher Education and Research in International Cooperation for Sustainability in 2023 displayed the paramount importance of think tanks such as the SDG Graduate School Alliance for future innovations. The conference provided highly relevant discussions on how to utilise the full potential of digital means in international education and research to promote the SDGs.

This volume is further testament to the Alliance's successful collaboration and a steppingstone for its future growth and commitment. The contributions offer valuable insights into the implementation and management of transnational academic partnerships. They highlight the challenges associated with such collaborations and provide practical examples of how to overcome them. By sharing their experiences, the authors of this book contribute to the growing body of knowledge on international higher education cooperation and inspire future generations of scholars and practitioners.

The findings presented by the authors are highly relevant for the DAAD as a learning organisation whose aim is to continuously improve its funding programmes. We are optimistic that this book will serve as a catalyst for further action and collaboration.

Dr. Kai Sicks
General Secretary
German Academic Exchange Service
(DAAD)
Bonn, Germany

Contents

Contributors

Tesfaye Abebe Amdie School of Plant and Horticultural Sciences, College of Agriculture, Hawassa University, Hawassa, Ethiopia

Sabina Appiah-Boateng School for Development Studies, University of Cape Coast, Cape Coast, Ghana

Dennis Avilés-Irahola Center for Development Research (ZEF), University of Bonn, Bonn, Germany

Christian Brandt Institute of Farm Management, University of Hohenheim, Stuttgart, Germany

Lea Caanitz Freie Universität Berlin, ZI Lateinamerika-Institut, Berlin, Germany

Laura Calderón Institute for Environmental Studies (IDEA), National University of Colombia, Bogotá, Colombia

Lucas-Andrés Elsner Habitat Unit, Technische Universität Berlin, Berlin, Germany;
Sustainable Cities and Climate Change, HafenCity Universität Hamburg, Hamburg, Germany

Emilia Fernengel TRA Present Pasts (TRA5), University of Bonn, Bonn, Germany

Lea Frauenknecht Institute for Cultural Policy, University of Hildesheim, Hildesheim, Germany

Marcel Frese Department of Chemistry, Organic and Bioorganic Chemistry, Bielefeld University, Bielefeld, Germany

Naomi Andrew Haruna Department of Industrial Studies, Faculty of Environmental Studies, University of Maiduguri, Maiduguri, Nigeria

Dirk Hollmann Interdisciplinary Faculty, Department Life, Light and Matter and Institute of Chemistry, University of Rostock, Rostock, Germany

Marie Huchzermeyer School of Architecture and Planning, University of the Witwatersrand, Johannesburg, South Africa

Udo Kragl Interdisciplinary Faculty, Department Life, Light and Matter and Institute of Chemistry, University of Rostock, Rostock, Germany

Taibat Lawanson Centre for Housing and Sustainable Development, University of Lagos, Lagos, Nigeria

Bruno Lenta Ndjakou Department of Chemistry, Higher Teacher Training College, University of Yaoundé I, Yaoundé, Cameroon

Esteban Mejía Leibniz Institute for Catalysis (LIKAT), Rostock, Germany

Sintayehu Yigrem Mersha School of Animal and Range Sciences, College of Agriculture, Hawassa University, Hawassa, Ethiopia

Aloysius Ngefac Department of English, Higher Teacher Training College, University of Yaoundé I, Yaoundé, Cameroon

Christian Petersheim Center for Development Research (ZEF), University of Bonn, Bonn, Germany

Carolina Tobón Ramírez Center for Development Research (ZEF), University of Bonn, Bonn, Germany

Frank Rasche International Institute of Tropical Agriculture (IITA), Nairobi, Kenya

Martin Ringsmut Department of Musicology, University of Vienna, Vienna, Austria

Rosario Rojas-Robles Institute for Environmental Studies (IDEA), National University of Colombia, Bogotá, Colombia

Bettina Schorr ZI Lateinamerika-Institut, Freie Universität Berlin, Berlin, Germany

Nicole Schönleber Institute of Farm Management, University of Hohenheim, Stuttgart, Germany

Norbert Sewald Department of Chemistry, Organic and Bioorganic Chemistry, Bielefeld University, Bielefeld, Germany

Tomás León Sicard Institute for Environmental Studies (IDEA), National University of Colombia, Bogotá, Colombia

Mfaniseni Fana Sihlongonyane School of Architecture and Planning, University of the Witwatersrand, Johannesburg, South Africa

Avhatakali Sithagu School of Architecture and Planning, University of the Witwatersrand, Johannesburg, South Africa

Le Thanh Son Faculty of Chemistry, VNU University of Science, Hanoi, Vietnam

Le Minh Thang Department of Chemical Engineering, School of Chemistry and Life Sciences, Hanoi University of Science and Technology, Hanoi, Vietnam

Tabea Angela Thiel Leibniz Institute for Catalysis (LIKAT), Rostock, Germany

Eva Youkhana Center for Development Research (ZEF), University of Bonn, Bonn, Germany

Abbreviations

ABS	Access and Benefit Sharing
ACS	American Chemical Society
AI	Artificial Intelligence
AR	Augmented Reality
BIBB	Federal Institute for Vocational Education and Training
BICC	Bonn International Center for Conflict Studies
BMBF	German Federal Ministry of Education and Research
BMZ	German Federal Ministry for Economic Cooperation and Development
CAPAZ	Colombian Institute for Peace
CARE	Collective Benefit, Authority to Control, Responsibility, Ethics
CASCAPE	Capacity Building for Scaling up of Best Practices in Agriculture in Ethiopia
CHSD	Centre for Housing and Sustainable Development
CITES	Convention on International Trade in Endangered Species of Wild Fauna and Flora
CLIFOOD	Climate Change Effects on Food Security
COLPAZ	Colombia Paz (activist group)
CPC	Centre Pasteur
CSPCS	Centre for the Study and Promotion of Cultural Sustainability
CSR	Corporate Social Responsibility
DAAD	German Academic Exchange Service
DFG	German Research Foundation
DSSP	Doctoral Studies Support Program on Environmental Peace and Development in Colombia
EN	English (or Entry Noted)
FAIR	Findable, Accessible, Interoperable, Reusable
FGR	Federal German Republic
FMN	Faculty of Mathematics and Natural Sciences
FSC	Food Security Centre
GDP	Gross Domestic Product

GDR	German Democratic Republic
GLP	Good Laboratory Practice
GS	Graduate School (or Global Sustainability)
HGF	Helmholtz Association of German Research Centers
HIC	Habitat International Coalition
HU	Hawassa University
HUST	Hanoi University of Science and Technology
ICT	Information and Communication Technologies
IDEA	Instituto de Estudios Ambientales
IMPM	Institute of Medical Research and Studies of Medicinal Plants
IT	Information Technology
IUCN	International Union for Conservation of Nature
LAI	Institute for Latin American Studies
LANACOME	National Drugs Quality Control and Validation Laboratory
LASA	Latin American Studies Association
LIKAT	Leibniz Institute for Catalysis
LLM	Master of Laws
MAT	Mutually Agreed Terms
MOOC	Massive Open Online Course
MOST	Ministry of Science and Technology
MPG	Max Planck Society
MUS	Master of Urban Studies
NAFOSTED	National Foundation for Science and Technology Development
NIVT	Vietnamese Institute for Vocational Training
NMR	Nuclear Magnetic Resonance
NTFP	Non-Timber Forest Products
NUA	New Urban Agenda
OCA	Environmental Conflict Observatory
ODI	Overseas Development Institute
PARG	Plans for Action and Regional Governance
PATR	Action Plans for Territorial Reform
PIC	Prior Informed Consent
REALISE	Realising Sustainable Agricultural Livelihood Security in Ethiopia
RORO	Recycling Organization for Research Opportunities
RPU	Peruvian Network of Universities
SALGA	South African Local Government Association
SAQA	South African Qualifications Authority
SIAMI	Environmental Information System
SNLMT	Santuario Nacional Los Manglares de Tumbes
SOAP	School of Architecture and Planning
UCL	University College London
UDS	University for Development Studies
UHOH	University of Hohenheim
UM	Urban Management
UNAL	National University of Colombia

UNILAG	University of Lagos
UR	University of Rostock
USA	United States of America
VAST	Vietnam Academy of Science and Technology
VGU	Vietnamese-German University
VR	Virtual Reality
WGBU	German Advisory Council on Global Change
ZALF	Leibniz Centre for Agricultural Landscape Research
ZEF	Center for Development Research
ZIDRES	Interest Zones for Rural and Economic Development
ZRC	Peasant Reserve Zones

Introduction

Christian Brandt, Esteban Mejía, and Bettina Schorr

Abstract This chapter introduces the objectives, scope, and context of the book, which explores the role of higher education in advancing the Sustainable Development Goals (SDGs) through international academic cooperation. Focusing on the DAAD Bilateral SDG Graduate Schools, it highlights how universities from the Global North and South collaborate to foster sustainable development, interdisciplinary knowledge production, and equitable partnerships. The chapter situates the initiative within the broader framework of the 2030 Agenda and discusses the transformative role of higher education institutions as agents of change. It underscores the need to bridge knowledge gaps on managing international academic collaborations and offers practical insights, best practices, and institutional experiences from diverse global contexts. Ultimately, it sets the stage for a detailed exploration of how higher education can drive global sustainability and innovation through inclusive, collaborative, and resilient approaches.

In 2015, the United Nations published the 2030 Agenda, the global development strategy following the Millenium Development Goals, which introduced a set of 17 Sustainable Development Goals (SDGs) to be achieved until 2030. This new framework represents several innovations compared to the previous strategy. The SDGs are much broader in scope, addressing not only essential poverty reduction goals but also a wide range of global challenges such as education, inequality, environmental sustainability, and peace. They are global, meaning they are relevant and applicable to all countries, regardless of their level of income and wealth. The SDGs are also

C. Brandt (✉)
Institute of Farm Management, University of Hohenheim, Stuttgart, Germany
e-mail: christian.brandt@uni-hohenheim.de

E. Mejía
Leibniz Institute for Catalysis (LIKAT), Rostock, Germany
e-mail: esteban.mejia@catalysis.de

B. Schorr
ZI Lateinamerika-Institut, Freie Universität Berlin, Berlin, Germany
e-mail: bettina.schorr@fu-berlin.de

C. Brandt et al. (eds.), *Implementing Sustainability and Internationalization in Higher Education*, World Sustainability Series, https://doi.org/10.1007/978-3-031-97837-1_1

indicator-based and require an empirical approach to measuring progress. A participatory approach is central to the SDGs, where governments are encouraged to define their primary goals through participatory processes involving local communities, civil society, and other stakeholders. Furthermore, the SDGs emphasize a subnational perspective, recognizing the importance of addressing issues not only at the national level but also at the local or regional level.

Several goals in the 2030 Agenda are new, including reducing inequality (SDG 10) and promoting education (SDG 4), specifically focusing on higher education. Higher education plays a crucial role not only in fostering social mobility but also as a key driver of sustainable development cooperation. Universities generate critical knowledge across disciplines, addressing sustainability challenges while training individuals to become change agents in their communities and professions. Additionally, they facilitate international networks and exchange opportunities, fostering global collaboration. This book explores the unique contributions of higher education to sustainable development, highlighting its role in achieving the SDGs.

Universities are increasingly seen as "living labs" where sustainability innovations can be tested and refined. Beyond research, universities contribute to sustainability by integrating sustainability principles into their operations, curricula, and community engagement activities. Research (among many others: Velasquez et al. 2006; Zilahy et al. 2009; Radinger-Peer and Pflitsch 2017; Withycombe et al. 2018; Leal Filho et al. 2020; Schorr et al. 2021) has shown that universities are essential promoters of sustainability, not only through academic research but also by embedding sustainability practices into day-to-day activities and fostering collaboration with local communities, businesses, and policy-makers. Universities' internationalization and transnational cooperation have further expanded their ability to contribute to global sustainability efforts.

The growing trend of academic cooperation, especially between universities in the Global North and South, has opened new opportunities for mutual learning, resource pooling, and intercultural exchange, which are crucial for addressing sustainability challenges in diverse contexts. Internationalization and transnational cooperation among universities have recently emerged as key drivers in sustainability efforts. This international cooperation helps promote sustainability by disseminating relevant knowledge, facilitating mutual learning, enhancing intercultural sensitivity, and pooling scarce resources.

1 The DAAD Bilateral SDG Graduate Schools Initiative

One initiative that connects universities to this international approach is the DAAD's funding line for SDG Graduate Schools, through which seven international SDG Graduate Schools received funding. The funding for this initiative is provided by the German Federal Ministry for Economic Cooperation and Development (BMZ). These SDG Graduate Schools aim to create an infrastructure for cooperative efforts between German higher education institutions and their counterparts in the Global

South. This initiative is in alignment with Agenda 2030, which calls for the global community to form partnerships to achieve the SDGs.

Achieving the SDGs requires multidimensional transformations grounded in systematic knowledge. Higher education institutions play a central role in this transformation by engaging in fundamental research and generating knowledge that can contribute to ecological, social, and economic sustainability. Universities also train young people who will become future decision-makers and practitioners in various sectors (government, business, non-profit organizations, etc.), empowering them to promote the sustainable well-being of their societies.

The seven DAAD Bilateral SDG Graduate Schools bring together universities from Germany, Latin America, Asia, and Africa, reflecting the multifaceted nature of Agenda 2030. These schools focus on capacity-building by developing research and teaching programs related to the different SDGs in the natural and social sciences. They promote academic internationalization and interdisciplinarity by encouraging exchanges among partners at all stages of development, which enables researchers and students to view global challenges from different regional perspectives. This allows for comparative and context-specific knowledge production, highlighting the interdependencies between different world regions. Furthermore, the SDG Graduate Schools engage with other epistemic communities and key stakeholders from the political, economic, and civil sectors to disseminate their findings and contribute to sustainable development.

2 Filling the Knowledge Gap on International Academic Cooperation

While the recognition of the importance of international academic cooperation for sustainability has grown, there is limited scholarly discussion on how these projects are conceptualized, established, and managed. This gap primarily arises from the temporary nature of many international collaborations, with projects often dissolving once the funding period ends, and valuable knowledge and experience being lost. Staff members frequently move on to other projects, taking with them the institutional memory of the collaboration. As a result, new project managers have few scholarly and organizational resources to draw upon when initiating new international cooperation projects in higher education.

This book seeks to fill this gap by providing institutional memory on the establishment and management of seven different international graduate schools worldwide. It offers insights into how these projects have been set up, implemented, and managed, focusing on the challenges faced and providing solutions and ideas for overcoming these obstacles. The book also presents a detailed description of the resources and instruments used to promote sustainable development through transnational academic cooperation.

The chapters of this book address key questions regarding the establishment and implementation of international cooperation projects aimed at promoting sustainability:

1. How can international projects (such as graduate schools) be built to promote sustainability?
2. What challenges do these types of international cooperation face, and what lessons can be learned from the experience?
3. How do bilateral graduate schools contribute to achieving the SDGs?

In particular, the book emphasizes that international cooperation in higher education is a dynamic process, requiring ongoing adjustments due to changing contextual factors, such as political conditions, budget constraints, administrative rules, and structures at partner universities. It provides rich insights into the processes of building, managing, and running these projects, showcasing the hands-on experience and accumulated knowledge of the contributors, who come from diverse perspectives, interests, and disciplinary backgrounds.

By offering lessons learned from the establishment and implementation of these bilateral SDG Graduate Schools, the book aims to accelerate knowledge gains in the development of future international cooperation projects. It helps to identify common challenges early on and avoid repeating mistakes. The chapters share good practices and showcase the impact of the programs in preparing future leaders dedicated to achieving the 2030 Agenda. The lessons shared in this book are valuable for anyone seeking to establish or manage transnational academic collaborations, with a particular focus on how such projects can contribute to the achievement of the SDGs. In the following, we resume the key insights of the different chapters, regarding knowledge production and interdisciplinarity, eye-level partnerships, tools for collaboration, learning, and impact, sustainable impact and the various challenges of international academic collaboration.

3 Knowledge Production and Interdisciplinarity

The SDG Graduate Schools have emerged as vital platforms for advancing multidisciplinary and interdisciplinary knowledge on sustainability, fostering innovative research that bridges diverse fields and disciplines. While achieving fully integrated interdisciplinarity remains a challenge, these initiatives have established collaborative spaces for cross-teaching, joint publications, and doctoral programs, which contribute to more comprehensive approaches to sustainability challenges.

In the Andean region, the *TrAndeS Graduate School* in Peru focuses on the complex relationship between social inequalities and sustainable development. Its research highlights historical and contemporary disparities shaped by ethnicity, class, gender, and space while addressing key sustainability issues such as extractivism, environmental degradation, and resource governance. By combining social science frameworks with sustainability insights, TrAndeS produces impactful knowledge through books, policy briefs, teaching materials, and research outputs that inform regional and global discourses.

Similarly, the *Doctoral Studies Support Programme (DSSP)* in Colombia integrates environment, conflict, and rural development through interdisciplinary and transdisciplinary research. By developing participatory methodologies that include local knowledge, DSSP creates innovative tools such as the Environmental Information System (SIAMI) and the Environmental Conflict Observatory (OCA). Its studies on environmental governance, agroecology, and alternative rural development models deliver practical insights for academia, policymakers, and communities.

In Africa, the *Performing Sustainability Graduate School* emphasizes cultural sustainability by exploring the intersections of culture, development, and peacebuilding in Nigeria and Ghana. Through the integration of arts and performance as research tools, the program challenges dominant Western frameworks, promoting inclusive and grassroots methodologies that respect local cultural systems. Research produced through fieldwork fosters critical and context-sensitive approaches to sustainability.

The *Wits-TUB-UNILAG Urban Lab* in South Africa addresses sustainable urban development in African cities by combining theory and practice. Its interdisciplinary programs generate knowledge on urbanization, informality, infrastructure deficits, and governance. Outputs include research on land governance, housing, and urban resilience, while summer schools and fieldwork bridge academic and real-world urban challenges, providing localized solutions informed by data-driven approaches.

In Southeast Asia, the *RoHan Graduate School* advances research on catalysis for sustainable resource management in Vietnam. By integrating chemistry, biology, and environmental sciences, RoHan contributes to clean water technologies, energy efficiency, and environmental sustainability. Research focuses on innovative methods such as photocatalysis for water purification and biomass conversion, leading to over 80 peer-reviewed publications and the establishment of modern catalysis laboratories.

In Cameroon, the *YaBiNaPA Graduate School* focuses on phytomedicine as a sustainable solution for health challenges in tropical regions. Using ethnopharmacological approaches, YaBiNaPA identifies, tests, and validates medicinal plants targeting parasitic and microbial diseases. The program develops standardized herbal formulations, such as malaria syrups and antifungal creams, and creates platforms for collaboration across chemistry, biology, and pharmacology, with significant outputs including over 1,000 identified secondary metabolites undergoing formulation testing.

In the context of food security and climate variability, the *CLIFOOD Graduate School* produces multidisciplinary knowledge by integrating agriculture, nutrition, and environmental sciences in Ethiopia. Through structured block seminars, joint teaching programs, and researcher exchanges, CLIFOOD fosters innovative insights into the linkages between climate change and food systems, addressing critical sustainability challenges.

4 Eye-Level Partnerships—A Shared Commitment to Equity and Innovation

The SDG Graduate Schools are united by their commitment to fostering eye-level partnerships between academic institutions in the Global North and South. These partnerships aim to create balanced and sustainable collaborations, allowing for critical reflections on inequalities within the global scientific community. While each program operates within unique thematic and regional contexts, they share a strong emphasis on mutual learning, local ownership, and South–South cooperation, addressing structural inequalities through inclusive academic engagement.

A common foundation across these initiatives is the development of trust-based relationships. This trust emerges through joint decision-making, transparent communication, and shared academic goals, ensuring that each partner contributes equally to the collaboration. The partnerships also go beyond academia, extending to local communities, policymakers, and regional networks, fostering meaningful engagement and knowledge exchange that reflects the realities of the Global South.

Despite their shared goals, the SDG Graduate Schools exhibit diverse approaches to fostering equity in their partnerships. The *TrAndeS Graduate School*, for example, emphasizes its collaboration between Freie Universität Berlin and Pontificia Universidad Católica del Perú as a model of long-standing trust and shared academic interests. TrAndeS strengthens institutional ownership by embedding its curriculum within PUCP's social sciences programs, ensuring sustainability and impact beyond the funding period. The program also extends its reach to regional public universities across Bolivia, Ecuador, Chile, and Peru, promoting capacity building while striving to balance academic excellence with representation of marginalized groups and voices often excluded from higher education.

Similarly, *RoHan* builds equitable and long-term partnerships between Germany's University of Rostock and Vietnam's Hanoi University of Science and Technology. The collaboration focuses on mutual capacity building, particularly through joint decision-making, co-supervised Master programs, and extensive researcher exchanges. By promoting leadership roles for Vietnamese scientists in academia and industry, RoHan ensures local ownership of the research process while its focus on environmental challenges, such as resource management and clean energy, addresses Vietnam's pressing national priorities.

The *YaBiNaPA Graduate School* highlights its partnership between Bielefeld University in Germany and the University of Yaoundé I in Cameroon as an example of integrating local and scientific expertise. The program places significant value on traditional knowledge systems by collaborating with indigenous healers, creating a space where ethnopharmacology and modern science converge. Importantly, YaBiNaPA includes researchers from less-favored regions of Cameroon and other African countries, such as Benin, Ethiopia, and Malawi, demonstrating its commitment to regional inclusion. Through its partnerships with policymakers and international advisory boards, the initiative ensures that research outcomes have both local and global relevance.

The *CLIFOOD* Graduate School underscores the value of equitable cooperation between the University of Hohenheim and Hawassa University in Ethiopia. This initiative combines North–South collaboration with South-South knowledge exchange, addressing inequalities in scientific knowledge production while strengthening local ownership. Partnerships are extended to regional stakeholders, ensuring meaningful engagement with food security and climate resilience challenges that are central to Ethiopia's sustainability agenda.

In Colombia, the *Doctoral Studies Support Programme (DSSP)* exemplifies a unique form of equity through its collaboration between the National University of Colombia and Germany's Center for Development Research (ZEF). The program integrates local and global perspectives to decolonize traditional academic hierarchies, creating space for voices that are often excluded from knowledge production. DSSP engages closely with grassroots organizations and local communities, ensuring that their lived experiences and insights remain central to the program's research agenda. This mutual learning process fosters a balanced exchange of academic and field expertise while promoting long-term social transformation.

The *Performing Sustainability Graduate School* builds partnerships that prioritize cultural and regional contexts in West Africa. Its collaboration between the University of Hildesheim in Germany, the University of Cape Coast in Ghana, and the University of Maiduguri in Nigeria highlights the importance of South-South cooperation alongside North–South exchanges. Through joint teaching programs, cotutelle arrangements, and resource-sharing, the program establishes balanced power dynamics that decolonize traditional partnerships.

In Africa's urban contexts, the *Wits-TUB-UNILAG Urban Lab* highlights the importance of equitable partnerships between the University of the Witwatersrand in South Africa, the Technical University of Berlin, and the University of Lagos in Nigeria. The program integrates local knowledge systems with global urban studies, addressing structural inequalities by prioritizing African urban challenges such as informality, infrastructure deficits, and governance. Equal participation in curriculum design, fieldwork, and teaching ensures that the collaboration remains both balanced and impactful. Alumni networks created through this program further strengthen ongoing cooperation across Sub-Saharan Africa.

While these initiatives share a strong commitment to equity and collaboration, they also face common challenges in navigating institutional differences, balancing academic excellence with inclusion, and overcoming power imbalances in global partnerships. Programs like TrAndeS and DSSP demonstrate that achieving equity requires addressing historical inequalities, decolonizing knowledge production, and promoting participation from marginalized groups. Initiatives such as RoHan and YaBiNaPA show how integrating local knowledge and scientific expertise can bridge gaps between traditional and modern systems. Meanwhile, Performing Sustainability and Wits-TUB-UNILAG highlight the importance of mutual respect, cultural sensitivity, and regional ownership in fostering sustainable and meaningful partnerships.

Overall, the SDG Graduate Schools demonstrate that equitable partnerships are not merely about sharing resources but about building inclusive, trust-based collaborations that respect local priorities, amplify marginalized voices, and contribute to sustainable development.

5 Creative tools for Collaboration, Learning, and Impact

The SDG Graduate Schools demonstrate a remarkable capacity for creativity and innovation in their approaches to teaching, research, international collaboration, and outreach. By employing a wide range of tools, from mobility programs and digital platforms to data management strategies and participatory methods, these initiatives provide practical solutions to global challenges while fostering international cooperation and capacity-building. Though unified by their commitment to innovation, each Graduate School tailors its methods to address specific regional and thematic needs, reflecting both shared principles and diverse approaches.

One significant innovation across the programs lies in the design of flexible mobility and exchange formats. For example, *RoHan* employs bilateral exchange programs that offer short- and long-term research stays in Germany and Vietnam, complemented by a Double-Degree Master Program in sustainable chemistry. This dual approach ensures both student mobility and structured curriculum development while leveraging advanced laboratory infrastructure to strengthen research capabilities. Similarly, *Performing Sustainability* focuses on mobility programs that include South-South exchanges between Ghana and Nigeria alongside research residencies in Germany, fostering cross-regional knowledge transfer while emphasizing arts and performance as tools for cultural sustainability.

Programs like *CLIFOOD* and *DSSP* adopt blended learning and digital tools to overcome geographical and institutional barriers. *CLIFOOD* combines on-site block seminars with ICT-based learning, creating innovative teaching formats that ensure continuity even during disruptions such as the COVID-19 pandemic and security challenges in Ethiopia. These efforts are further supported by a scientific data center that provides satellite-based weather and land-use data, enabling interdisciplinary research on food security and climate change. In contrast, *DSSP* uses virtual platforms like SIAMI to systematically compile and disseminate environmental data, ensuring accessibility for researchers and communities. Additionally, DSSP introduces participatory tools like ColPaMon, which involve local communities in tracking environmental conflicts, and organizes international symposia and workshops to bridge academic and non-academic perspectives, promoting dialogue and co-creation of knowledge.

The programs also demonstrate creative approaches to capacity-building and interdisciplinary teaching. For instance, *TrAndeS* introduces in-situ scholarships, allowing students to access high-quality education locally, thereby reducing academic brain drain. To support regional researchers, TrAndeS also provides postdoctoral writing fellowships, facilitating the production and dissemination of knowledge. Its interdisciplinary teaching methods combine anthropology, economics, sociology, and political science, enabling students to analyze the complexities of inequality and sustainability. These innovations are further complemented by the establishment of the Red trAndeS network, which fosters regional cooperation and knowledge exchange across the Andean region.

YaBiNaPA and *RoHan* distinguish themselves through their investment in cutting-edge research infrastructure. YaBiNaPA has established state-of-the-art laboratories for natural product analysis, enabling groundbreaking research on medicinal plants for parasitic and microbial diseases. The program further implements the Nagoya Protocol on Access and Benefit Sharing (ABS), ensuring ethical and sustainable use of natural resources. Through workshops, open days, and field studies, YaBiNaPA connects researchers, traditional healers, policymakers, and stakeholders, bridging the gap between modern science and indigenous knowledge systems.

The use of soft-skill training and integrated learning models is another key feature of these Graduate Schools. *Performing Sustainability* combines biannual workshops with online colloquia, offering students opportunities to develop methodological, theoretical, and professional skills. Similarly, *CLIFOOD* incorporates programs to improve soft skills like scientific writing, communication, and problem-solving, preparing researchers for academic and practical applications. These approaches are mirrored in RoHan's hybrid learning models, which include flipped classrooms, virtual seminars, and e-learning tools, increasing global participation and accessibility.

In conclusion, the SDG Graduate Schools embody innovation through their creative use of teaching formats, research tools, and collaborative platforms. From interdisciplinary teaching and data management systems to participatory tools and mobility programs, these initiatives reflect a shared commitment to building sustainable solutions through innovative practices. By tailoring their instruments to regional and thematic needs, each Graduate School showcases unique contributions while collectively advancing the frontiers of global research, education, and cooperation.

6 Lasting Legacies—Building Sustainable Impacts Beyond Funding

The SDG Graduate Schools have demonstrated their capacity to leave enduring legacies by creating lasting institutional structures, strengthening research capacity, and establishing networks that persist well beyond the funding phase. These initiatives ensure their impacts are sustained through alumni networks, and academic programs that continue to act as multipliers, fostering collaboration, knowledge dissemination, and capacity-building at local, regional, and global levels.

A significant shared outcome of the Graduate Schools is the institutionalization of programs that embed interdisciplinary and sustainable frameworks within partner universities. For example, *TrAndeS* has contributed to successfully institutionalize doctoral scholarships and postdoctoral fellowships at Pontificia Universidad Católica del Perú (PUCP), strengthening academic capacity in the Andean region. Its legacy extends to regional public universities across Bolivia, Ecuador, Chile, and Peru, where it fosters integration and reinforces research on inequality and sustainability. Similarly, *DSSP* in Colombia has created structured doctoral and postdoctoral programs that produce research with tangible relevance for policymakers, academia, and local communities. By institutionalizing tools such as the Environmental Information System (SIAMI) and the Environmental Conflict Observatory (OCA), DSSP ensures long-term environmental monitoring and integrated knowledge systems that support peacebuilding and sustainable development.

In Africa, *Wits-TUB-UNILAG Urban Lab* has institutionalized urban management degrees at the University of the Witwatersrand (Wits) and the University of Lagos (UNILAG), addressing critical gaps in African urban education. Through strengthened faculty development, IT infrastructure, and research programs, the Urban Lab contributes to sustainable urban practices. Similarly, *Performing Sustainability* has created long-lasting structures such as the Centre for the Study and Promotion of Cultural Sustainability (CSPCS) in Maiduguri. These institutional centers serve as hubs for regional collaboration and capacity-building, ensuring cultural sustainability remains a focus in academic and community-driven initiatives.

Programs like *RoHan* and *YaBiNaPA* have prioritized the creation of research hubs equipped with state-of-the-art facilities to advance scientific knowledge in their respective fields. *RoHan* has established the German-Vietnamese Catalysis Research Centre (GeViCat) as a regional leader for catalysis research in Southeast Asia, ensuring its continued contributions to resource management and clean energy innovations. Likewise, *YaBiNaPA* has enhanced research infrastructure at the University of Yaoundé I, positioning itself as a scientific hub for natural product research. These advancements include developing a Cameroonian pharmacopoeia, conserving biodiversity through sustainable plant use education, and producing herbal formulations to address health challenges.

An equally significant legacy across the SDG Graduate Schools lies in the strengthening of alumni networks as multipliers of knowledge and innovation. These networks ensure the continuity of collaboration, knowledge dissemination, and capacity-building within and beyond their regions. In *CLIFOOD*, alumni networks serve as central actors in fostering ongoing research partnerships and initiating locally funded PhD programs in SDGs, showcasing the scalability and replicability of the Graduate School model. Similarly, *RoHan* and *YaBiNaPA* alumni contribute to academia, industry, and government, taking on leadership roles that advance sustainability in their respective fields while promoting gender inclusivity.

The *Performing Sustainability* and *Wits-TUB-UNILAG Urban Lab* initiatives demonstrate the lasting influence of alumni networks in driving sustainability initiatives. Alumni from these programs continue to contribute to academic capacity-building, urban policy innovation, and cultural preservation, strengthening regional cooperation across West and Sub-Saharan Africa. Meanwhile, *TrAndeS* alumni and members of its long-term research network Red trAndeS act as agents of knowledge production, capacity-building, and policy influence across the Andean region, ensuring that research on social inequalities and sustainability remains a central focus.

Finally, these programs emphasize capacity-building as a critical element of their legacies. Together, the SDG Graduate Schools have trained hundreds of Ph.D. and postgraduate researchers, equipping them with advanced scientific tools and methodologies to address local and global challenges.

In conclusion, the SDG Graduate Schools have established lasting institutional frameworks, robust research infrastructure, and dynamic alumni networks that extend their impacts far beyond the funding period. These legacies ensure that the programs remain multipliers of knowledge, equity, and innovation, contributing to sustainable development and academic excellence across diverse global contexts. Through capacity-building, regional integration, and collaborative tools, the Graduate Schools demonstrate the transformative power of education and research in addressing global challenges.

7 Challenges—Navigating Barriers to International Cooperation

The SDG Graduate Schools grapple with a range of challenges inherent in international cooperation, stemming from institutional, cultural, and structural differences as well as broader global crises and local inequalities. While the challenges are shared across programs, each initiative highlights unique obstacles shaped by its regional focus and thematic priorities, offering tailored solutions that reflect adaptability and innovation.

A recurring challenge is the complexity of collaboration across diverse institutional and cultural settings. This is described vividly by one alumnus of *Performing Sustainability* as "*three different people, three different ways of doing things, but trying to make it work as one person. One voice, but different road maps*". Programs like *TrAndeS* and *RoHan* emphasize the need for trust-building, transparent communication, and flexible management strategies to overcome these differences. For TrAndeS, administrative and regulatory misalignments between German and Peruvian institutions required innovative solutions, while RoHan relied on collaborative management to address cultural and institutional differences between Germany and Vietnam. Similarly, *Wits-TUB-UNILAG* faced the challenge of aligning academic systems across Germany, South Africa, and Nigeria, which required the design of flexible learning formats and collaborative curriculum development.

Another shared challenge is resource and infrastructure limitations, particularly for programs based in regions with underdeveloped facilities or political instability. For instance, *CLIFOOD* and *YaBiNaPA* faced infrastructural issues such as weak internet, power outages, and insufficient research facilities. CLIFOOD's work in

Ethiopia was further disrupted by political unrest and conflicts, while *DSSP* in Colombia had to contend with ongoing security risks that hindered field research. Solutions involved enhancing technical infrastructure, creating blended learning models, and prioritizing resilience strategies such as virtual platforms for continuity during crises.

Programs like *YaBiNaPA* and *Performing Sustainability* highlight gender and regional inequalities as significant challenges. In YaBiNaPA, cultural and structural barriers limited women's participation in science, which was addressed through gender-sensitive recruitment and the creation of leadership opportunities for female scientists. Performing Sustainability also emphasized the societal pressures faced by female scholars, requiring additional support to balance academic aspirations with family expectations. Furthermore, YaBiNaPA tackled regional inequalities by supporting researchers from conflict-affected areas like Northwest Cameroon, ensuring equitable participation.

Power dynamics and the need to decolonize research emerged as critical challenges in programs such as *DSSP*, *Performing Sustainability*, and *Wits-TUB-UNILAG*. DSSP faced tensions arising from differing interpretations of key concepts like "empowerment" and "sustainability," highlighting the need for mutual respect and flexibility in research timelines to accommodate local processes. Similarly, Performing Sustainability navigated historical imbalances by fostering equitable partnerships through continuous dialogue and compromise, while Wits-TUB-UNILAG worked to balance local knowledge systems with global urban studies to address colonial legacies in education.

All programs encountered challenges related to global crises like the COVID-19 pandemic. These crises forced a rapid transition to digital teaching and learning formats. While disruptive, the shift spurred innovations in hybrid learning models and e-learning platforms, enabling the programs to maintain academic continuity and increase global accessibility.

Sustainability of funding remains a persistent challenge for all SDG Graduate Schools. *Performing Sustainability* noted the difficulty in securing long-term funding beyond DAAD support, highlighting the vulnerability of cultural research to financial constraints. *TrAndeS* and *DSSP* similarly faced funding limitations but addressed these challenges through low-cost networking activities, leveraging individual contributions, and building long-term alliances with local actors to sustain program impacts.

Finally, interdisciplinary tensions were a notable challenge for programs like *TrAndeS* and *DSSP*, which sought to bridge the gap between social and natural sciences. Achieving genuine interdisciplinarity required ongoing dialogue, collaboration, and flexibility to ensure that diverse perspectives were integrated effectively.

In conclusion, while the SDG Graduate Schools face common challenges such as institutional misalignments, resource constraints, gender inequalities, and power imbalances, their responses demonstrate remarkable adaptability and innovation.

By implementing solutions like trust-building strategies, virtual platforms, gender-sensitive approaches, and resilient management tools, these programs offer valuable insights into overcoming barriers to international cooperation. Their experiences underscore the importance of flexibility, mutual respect, and context-sensitive solutions in fostering equitable and sustainable academic partnerships.

The SDG Graduate Schools illustrate the transformative potential of higher education in advancing sustainability and promoting equitable international cooperation. Through interdisciplinary research, innovative partnerships and localised engagement, these initiatives have not only generated critical knowledge, but also empowered future leaders to drive meaningful change. Despite the challenges of navigating institutional differences, resource constraints and global crises, these programs have demonstrated resilience, adaptability and a commitment to decolonising academic collaboration. By highlighting good practices and lessons learned, this book serves as both a reflection on past achievements and a guide for future endeavours in higher education and sustainable development. As the world continues to grapple with complex environmental, social and economic challenges, the insights from these graduate schools offer valuable roadmaps for promoting inclusive, impactful and long-term solutions through education, research and international cooperation.

8 Organization of the Book

This volume presents the experiences, achievements, and challenges of the seven international SDG Graduate Schools supported by the German Academic Exchange Service (DAAD) and the Federal Ministry for Economic Cooperation and Development (BMZ). Each chapter offers a detailed account of how international academic partnerships have contributed to the Sustainable Development Goals (SDGs) through diverse thematic lenses and regional contexts. The schools combine interdisciplinary academic collaboration with practical capacity-building strategies, showcasing different pathways toward sustainability, equity, and global cooperation in higher education. Taken together, the chapters offer insights into the complex but vital terrain of international research partnerships and the role of graduate education in promoting the SDGs. Each chapter has been written by coordinators and principal investigators of the respective SDG Graduate School. The editors have respected the individual styles, perspectives, and methodological approaches of the authors, acknowledging the diversity of contexts and experiences represented in this volume. The book is organized as follows.

It opens with the German-Peruvian Graduate School "trAndeS—Social Inequalities and Sustainable Development in the Andean Region", jointly coordinated by Freie Universität Berlin and Pontificia Universidad Católica del Perú. This chapter examines how persistent inequalities—shaped by class, ethnicity, gender, and territory—intersect with sustainable development in the Andean region. Through interdisciplinary research, postgraduate education, and regional networking, trAndeS confronts the structural barriers to sustainable development, while also documenting the challenges of enhancing academic quality and institutionalizing international cooperation in this context.

Building on this analysis of inequality and development, the second chapter turns to DSSP—Environmental Peace and Development in Colombia, a bilateral partnership between the National University of Colombia (IDEA) and the University of Bonn (ZEF). Focused on the links between environment, conflict, and development, this program integrates inter- and transdisciplinary approaches to foster environmental peace. The chapter highlights the importance of local knowledge, participatory research, and academic reflection in rethinking higher education's contribution to the SDGs in a post-conflict context.

From Latin America to West Africa, the third chapter examines the trilateral Graduate School "Performing Sustainability: Cultures and Development in West Africa", operating between Germany, Ghana, and Nigeria. With an emphasis on gender equality, decolonial research methods, and community-based practices, this school explores cultural dimensions of sustainability. It stresses the value of grassroots engagement and critical interpretations of the SDGs, pointing to the complexities of knowledge exchange across postcolonial academic landscapes.

The theme of sustainable urban development is the focus of the next chapter, which discusses the Wits-TUB-UNILAG Urban Lab, a collaboration between universities in South Africa, Germany, and Nigeria. In response to Africa's rapid and uneven urbanization, the program has introduced new master's curricula in urban management. By analyzing urban policy discourses and historical inequalities in planning knowledge, this chapter reflects on how international academic partnerships can strengthen capacity for sustainable urban futures.

The volume then shifts to the field of natural sciences with the RoHan SDG Graduate School in Chemistry, linking the University of Rostock and Hanoi University of Science. What began as small-scale collaboration has evolved into a long-term strategic partnership centered on catalysis and sustainable chemistry. The chapter emphasizes the role of interdisciplinary research and practical applications in developing technological solutions for both local and global challenges.

Continuing with a scientific focus but grounded in traditional medicine, the sixth chapter explores the YaBiNaPA Graduate School between Bielefeld University and the University of Yaoundé 1. Here, research addresses urgent health and development issues in Cameroon, particularly through the study of natural products with antiparasitic and antibacterial properties. The chapter examines how this South-North partnership fosters knowledge transfer, supports SDG-related goals in health and biodiversity, and promotes the implementation of the Nagoya Protocol.

The final chapter focuses on food security and climate resilience in Ethiopia through the CLIFOOD Graduate School, coordinated by the University of Hohenheim and Hawassa University. Addressing climate change impacts on rain-fed agriculture, CLIFOOD supports interdisciplinary education and international cooperation to develop sustainable agricultural practices. The chapter reflects on lessons learned in project design and implementation, underscoring how bilateral graduate schools can strengthen institutional capacities and promote integrated sustainability solutions.

Together, these chapters provide a rich mosaic of international academic cooperation grounded in diverse disciplines and regional realities. They illuminate how the SDG Graduate Schools have not only advanced research and education but also nurtured enduring partnerships in the shared pursuit of sustainable development. As editors, we extend our sincere gratitude to all the colleagues from around the world who contributed to this volume and with whom we have had the privilege of working over the past decade.

References

Leal Filho W, Salvia AL, Pretorius RW, Brandli LL, Manolas E, Alves F, Azeiteiro U, Rogers J, Shiel C, Do Paco A (eds) (2020) Universities as living labs for sustainable development: supporting the implementation of the sustainable development goals. Springer, Cham

Radinger-Peer V, Pflitsch G (2017) The role of higher education institutions in regional transition paths towards sustainability. Rev Reg Res 37:161–187. https://doi.org/10.1007/s10037-017-0116-9

Schorr B, Braig M, Fritz B, Schütt B (2021) The Global knowledge value chain on sustainability: addressing fragmentations through international academic partnerships. Sustainability 13(17):9930–9950. https://doi.org/10.3390/su13179930

Velazquez L, Munguia N, Platt A, Taddei J (2006) Sustainable university: what can be the matter? J Clean Prod 14(9–11):810–819

Withycombe L, Beaudoin F, Lerner A, John B, Beecroft R, Tamm K, Wiek A, Lang D (2018) Transferring sustainability solutions across contexts through city-university partnerships. Sustainability 10(9):2966. https://doi.org/10.3390/su10092966

Zilahy G, Huisingh D, Melanen M, Phillips VD, Sheffy J (2009) Roles of academia in regional sustainability initiatives: outreach for a more sustainable future. J Clean Prod 17(12):1053–1056. https://doi.org/10.1016/j.jclepro.2009.03.006

Social Inequalities and Sustainable Development in the Andean Region: TrAndeS—A German–Peruvian SDG Graduate School

Bettina Schorr and Lea Caanitz

Abstract This chapter explores the intricate relationship between social inequalities and sustainable development in the Andean region, focusing on research and academic collaboration facilitated by the trAndeS program. Established by Freie Universität Berlin and Pontificia Universidad Católica del Perú, trAndeS aims to address the persistent and multidimensional nature of inequalities in the Andean countries, spanning class, ethnicity, gender, and territory. The program promotes interdisciplinary research and training, fostering a deeper understanding of how these inequalities affect opportunities for sustainable development. The analysis covers the historical evolution of regional inequalities, emphasizing the profound social, political, and environmental challenges that persist despite recent economic booms. Through postgraduate education, interdisciplinary research, and regional networking, trAndeS seeks to empower scholars and practitioners to mitigate these inequalities. The chapter highlights key research outputs, including the dynamics of extractive economies, governance, and environmental degradation, alongside the practical challenges encountered in establishing and institutionalizing such an international academic collaboration. Ultimately, the work underscores the importance of addressing social inequalities as a cross-cutting issue integral to achieving sustainable development in the Andean region.

1 Introduction

Social inequality has been a concern of the international development community since the 1980s. The UN-commissioned Brundtland Report (1987), which coined the term 'sustainable development', extensively discussed the detrimental effects of social inequality on people and the environment. The rise of global inequality and the publication of several highly acclaimed academic bestsellers, such as Piketty's "Capital in the 21st Century" (2014), has increased this attention over the last two

B. Schorr (✉) · L. Caanitz
Freie Universität Berlin, ZI Lateinamerika-Institut, Berlin, Germany
e-mail: bettina.schorr@fu-berlin.de

C. Brandt et al. (eds.), *Implementing Sustainability and Internationalization in Higher Education*, World Sustainability Series, https://doi.org/10.1007/978-3-031-97837-1_2

decades. In parallel, research on inequality shifted from a narrow economic perspective of income and (sometimes) wealth to a multidimensional notion. This perspective emphasizes that social inequalities can take multiple forms in different domains, including 'horizontal' group inequalities (Stewart 2008; Tilly 1998), unequal access to and power over political decisions and processes (Kreckel 2004), generally benefitting small groups of "power elites" (Mills 1956), and socio-ecological inequalities that shape people's opportunities to live in a healthy natural environment and to access natural resources (Góngora 2015; Göbel et al. 2014; Dietz 2017).

In 2015, reducing inequality became an official goal (SDG 10) included in the 17 Sustainable Development Goals (SDGs) of the UN 2030 Agenda. While this is a significant improvement over the previous Millennium Development Goals strategy (2000–2015, cf. Fukuda-Parr 2016; Freisstein and Mahlert 2016; Melamed 2012), it is misleading to think of inequality as an isolated goal among others: Social inequality is a thoroughly cross-cutting force that affects all dimensions of sustainable development (Schorr 2018). For instance, within the group of 17 SDGs, the development problems addressed in SDG 1 and 2 on reducing poverty and hunger, as well as the ones addressed in SDG 3 on improving access to health care, SDG 4 on quality education, and SDG 6 on clean water and sanitation are often not the result of resource scarcity or inefficient allocation. Instead, the deliberately created unequal distribution of assets and resources leads to poorer groups or individuals falling behind. Furthermore, the goal of gender equality (SDG 5) reflects the growing exclusion and discrimination against women, which enormously hinders opportunities for sustainable development worldwide.

Latin America stands out as one of the most unequal regions in the world, paralleled at times only by Sub-Saharan Africa. While a marginal subject in the past, recent research has extensively documented the dimensions and consequences of social inequality in Latin America (cf., amongst others, Therborn 2006, 2013; López-Calva and Lustig 2010; Reygadas 2008, 2015; Gootenberg and Reygadas 2010; Costa 2011; Gasparini and Cruces 2013; Braig et al. 2015; Blofield 2015; Boatcă 2015; Jiménez 2015; Fritz and Lavinas 2015; Castillo and Maldonado 2015; Atria et al. 2018; Jelin et al. 2017; Sánchez-Ancochea 2021; Maldonado and Schorr 2023). These works show that more than in almost any other region of the world, social inequalities are multidimensional and structured along intersecting lines of class, ethnicity, gender, age, and territory that mutually reinforce each other. Moreover, after focusing extensively on the poor and excluded, recent scholarship has shed light on the role of the wealthy in producing and perpetuating social inequalities (Heredia 2023).

However, research in this line has concentrated on the larger Latin American countries, Mexico and Brazil (Amarante et al. 2019). The Andean region, except for Chile, has not been the center of academic research on inequality, although inequalities in this region are profound, multifaceted, and intrinsically interlinked. In addition, Andean countries' higher education sectors are generally weak (Schorr et al. 2021). Universities focus on teaching and rarely produce or promote original, evidence-based research, particularly in the social sciences. These constraints in higher education and research have also limited knowledge production in the field of inequality studies in the area.

Against this backdrop, Freie Universität Berlin (FU Berlin) and the Pontificia Universidad Católica del Perú (PUCP, based in Lima, Peru) established in 2016 the combined training and research program "trAndeS—Advanced Studies on Social Inequalities and Sustainable Development". The program was facilitated by a grant from the German Academic Exchange Service (DAAD) and the German Ministry for Economic Cooperation and Development.

trAndeS focuses on the Andean region, comprising, in a broad sense, Argentina, Bolivia, Chile, Colombia, Ecuador, Peru and Venezuela. The acronym 'trAndeS' reflects the program's defining features: a regional focus on the Andean region (*Andes*), a thematic emphasis on sustainability and social inequality (the "*S*"), and the goal of fostering a transversal, cross-cutting dialogue that spans countries, disciplines, and stakeholder groups ("*Trans*"). trAndeS pursues two overall objectives: First, to promote a research agenda on the impact of social inequalities on sustainable development in Andean countries, identifying the actors that cause or suffer from inequalities and the mechanisms that (re)produce them. Second, it aims to strengthen research and teaching capacities at partnering institutions in the Andean region. These objectives are addressed through a three-pillar work program: The first pillar is postgraduate training, including a scholarship program for Masters students (until 2021) and doctoral students (until 2026). In addition to their regular studies in selected disciplinary programs of the Department of Social Sciences at the PUCP, trAndeS students complete a complementary curriculum of two courses introducing critical concepts and theories from inequality studies and providing an empirical overview of social inequalities in the Andean region, including their historical evolution. The curriculum also requires trades students to take academic "soft skills" courses (such as "academic writing" and "project management"), to discuss the progress of their research in international colloquia, and to participate in extracurricular activities such as international conferences and workshops.

The second pillar is interdisciplinary research. It includes the research projects of the scholarship holders and the postdoctoral fellows. Moreover, five thematic working groups were established: gender, power asymmetries and governance, economic volatility and resource economies, environmental degradation and social inequality, and green states and societies.[1] In these groups, senior and junior researchers from the partner universities and other Andean institutions work together to produce and transfer academic knowledge.

The third pillar is networking. In addition to strengthening the ties between the PUCP and the FU Berlin, the most important components of the networking strategy are the trans-Andean network (Red trAndeS) and transdisciplinary outreach activities with the development and public sectors. Within the Red trAndeS, which includes researchers and institutions from all Andean countries, most of the capacity-building efforts with other regional universities take place.

trAndeS is directed by an Executive Board of four members from each partner, responsible for strategic questions, monitoring the achievement of the objectives, and quality management. A binational directorate appointed by the Executive Board

[1] For more information on the working groups, see https://www.programa-trandes.net/index.html.

oversees the operational management of the project, and two project offices located at the PUCP and the FU Berlin provide administrative and technical assistance to the program.

This chapter is divided into two parts. The first part reviews the trAndeS research program and its principal research outputs, focusing on the historical evolution of multidimensional social inequalities in the region and the inequality-related dynamics during the most recent economic cycle from around 2000 to 2020. The second part discusses the experience of conducting the trAndeS program and its primary outcomes. It starts by highlighting lessons learned in establishing and managing the international program. The subsequent section reflects on the program's achievements and challenges in four key areas: the scholarship program, interdisciplinary training and research, networking, and institutionalization. The chapter concludes by emphasizing the importance of international academic cooperation in promoting transitions toward sustainable societies.

2 Social Inequalities in the Andean Region and How They Affect Sustainable Development

The research agenda of the trAndeS program focuses on the relationship between social inequalities and sustainable development in the Andean region. It revolves around the core concept of "multidimensional, interdependent inequalities", defined as "the distances between positions which individuals or groups of individuals assume in the context of a hierarchically organized access to relevant social goods (income, wealth, a healthy environment, etc.) and power resources (rights, political participation, political positions, etc.)" (Jelin et al. 2017: 1). Social inequalities are always relational and interdependent, connecting at least two actors who can be located at different political levels or spaces from the local to the global.

This section offers an overview of how social inequalities impact opportunities for sustainable development in the Andean region. It begins with a historically informed review of the region's emergence and evolution of social inequalities. Next, it concentrates on the period from 2000 to 2020, the program's focal period. This period saw an exceptional economic boom followed by a downturn, significantly influencing inequality dynamics and the region's sustainable development opportunities. The discussion is based on original research conducted within trAndeS (by scholarship holders and principal investigators, marked **in bold**) and academic secondary literature, especially from researchers participating in the Red trAndeS.

Social Inequalities in the Andes: Deeply rooted, Layered and Fluid

Multidimensional inequalities are pervasive in Andean countries. Although these inequalities have deep historical roots, they have never been static and evolved over time. As a result, Andean countries today exhibit multiple "layers" of inequalities

(Baquero-Melo 2014), which were introduced or modified at various historical points (**Schorr** 2023; **Henríquez and Vila** 2023).

Social inequalities in Andean countries can be traced back to Spanish colonization. Beginning in the sixteenth century, Spanish rule subjugated Indigenous communities, particularly in the Andean highlands, creating deep ethnic and horizontal inequalities through a system of social segregation. Enslaved Africans were also brought to the region, forced to work in mines and on plantations, becoming part of a skin-color-based inequality regime (Góngora et al. 2019). Colonial rule further established a significant concentration of wealth and power in the hands of a few families, initially of European descent.

With the end of colonial rule in the early nineteenth century, power dynamics shifted toward mestizo (people of mixed European and Indigenous ancestry) and creole (descendants of Europeans born in the colonies) societies. Although these changes altered the social landscape, they did not significantly improve conditions for most Indigenous and black peoples. Instead, they compounded existing social inequalities with new asymmetries. In the nineteenth century, feudal rule in the countryside persisted in large "haciendas" (extensive estates or plantations), subjugating Indigenous people in slave-like conditions to the owners' power and discretion. When slavery and feudalism were abolished, the electoral rights in the new republics remained restricted to affluent white males, excluding indigenous and rural populations as well as women. Creole and mestizo families accumulated (further) wealth through businesses typically related to their countries' natural resources (the "tin barons" in Bolivia and the "guano entrepreneurs" in Peru). Over time, many of these families evolved into corporate dynasties, owning large, multipurpose economic conglomerates that combine services with commerce and production (Ross Schneider 2008; Monsalve Zanatti 2014; Durand 2013). The "discovery" of the resource-rich Amazon region initially attracted migrants seeking to exploit its rubber resources, leading to significant human rights violations against Indigenous Amazonian communities and contributing to the degradation of the ecosystem (**Damonte and Schorr** 2022).

Urban expansion led to the rise of a marginalized poor working class, many of Indigenous descent, intertwining ethnic discrimination with class prejudices. While development efforts concentrated on metropolitan areas, vast territories within the new nation-states were excluded from public investments and services, exacerbating territorial or spatial inequalities (**Just 2019**). Mining sites proliferated across the Andes, where Indigenous miners worked under harsh and inhumane conditions (Nash 1979). Due to the absence of meaningful rural development, the migration from rural to urban areas further widened the gap between different societal strata and regions.

In the latter half of the twentieth century, various socio-political shocks exacerbated territorial and spatial stratification, deepening social inequalities. Military dictatorships, beginning in the 1960s, waged violent internal campaigns against opposition forces, resulting in widespread human rights violations and forcing thousands into exile. During the 1980s, in the context of democratization, governments implemented "neoliberal" adjustment policies promoted by international organizations such as the World Bank to address economic crises. These policies, aimed at reducing the state's role, led to the closure of numerous state-owned enterprises and significant cuts in public investments, including social welfare programs. As a result, unemployment, economic informality, and the urban poor population rose sharply.

During the twentieth century, Andean countries also experienced several critical political junctures that activated inclusive forces with the potential to level the unequal playing field. These include land reforms, in some countries in the frame of full-fledged national revolutions (e.g. in Bolivia and Chile) during the 1950s and 60s, and the transition to formal democracy that ended military rule throughout the region by 1990 (except in Peru, where dictator Alberto Fujimori, known for his authoritarian practices and human rights abuses, remained in power until 2000). Electoral and decentralization reforms during the 1990s aimed to address territorial inequalities by strengthening subnational political units and improving democratic representation. In parallel, increased globalization and transnationalism fostered alliances on social justice issues and increased cooperation with external actors. Bi- and multilateral organizations allied with national and international NGOs to address social inequalities' most significant ramifications by providing funds, expertise, networks and communication channels. Overall, citizen participation and activism increased with the advent of democracy. Especially Indigenous activism fostered the adoption of multicultural policies and the recognition of Indigenous rights in the Andes, featuring significant portions of Indigenous populations (Van Cott 2000).

The New Century: Boom, Bust, and the Evolution of Social Inequalities

Since the sixteenth century, colonialism in Andean countries created deep ethnic, racial, and economic inequalities. After colonialism, mestizo and creole elites maintained these inequalities through feudalism and political dominance. Urbanization, unequal infrastructure investment, extractivism, and political instability further widened social and regional gaps. Although democratization and globalization introduced significant improvement, the layered inequalities persisted into the New Century.

In the early 2000s, a global commodity boom, driven largely by Chinese demand, became the most significant "leveling force" affecting social inequalities in the Andean region. Lasting until the mid-2010s, this boom pushed prices for Andean commodities to unprecedented highs (Bebbington 2011). As a result, governments across the region embraced extractive development models focused on extracting and exporting raw materials, reflecting a widespread and politically diverse "commodity consensus" (Svampa 2015). Andean countries saw a substantial expansion of their extractive sectors, including traditional areas like mining and hydrocarbon extraction, and agrarian sectors. This new "agrarian extractivism" (McKay et al. 2021) involved the production of non-traditional crops such as soybeans (in Bolivia and Argentina), avocados (in Chile and Peru), and asparagus (in Peru), alongside traditional crops like quinoa, an Andean grain increasingly sought after by health-conscious consumers in Global North countries. Concurrently, the economic boom coincided with a leftward political shift (the "pink tide"), bringing in various countries presidents and coalitions committed to social inclusion and redistributive policies to power (Cameron and Hershberg 2010).

This recent economic surge has profoundly shaped the socio-economic landscape of the Andean region, significantly impacting opportunities for sustainable development, both positively and negatively. With an influx of rents and capital, all countries notably reduced poverty and expanded their middle classes (Stampini et al. 2016). Income inequality, previously among the highest globally, saw a sharp decline (López-Calva and Lustig 2010; ECLAC 2019), prompting scholars to speak of a "historic moment of equality" for Latin America (Fritz and Lavinas 2015). The GINI coefficient for income distribution, a standard measure of income inequality, dramatically decreased across Andean countries: in Argentina from 0.47 to 0.39, Bolivia from 0.61 to 0.45, Chile from 0.51 to 0.45, Colombia from 0.57 to 0.51, Ecuador from 0.54 to 0.44, and Venezuela from 0.42 to 0.38 (ECLAC 2019).

However, the improvements in economic inequality have not been accompanied by a reduction in other forms of social inequality, underscoring the importance of a multidimensional understanding of social inequalities. Horizontal and categorical inequalities remain significant in the region. Indigenous populations, especially women, lag behind in access to public services such as education, water, sewage, and healthcare (Hall and Patrinos 2012; ECLAC 2019). Despite the introduction of formal democracy, various participatory mechanisms, and specific (Indigenous) rights, political exclusion persists, with marginalized groups encountering obstacles to political engagement and democratic representation (Schilling-Vacaflor and Flemming 2015). Efforts toward judicial reforms to recognize Indigenous justice systems have not led to genuine judicial pluralism. Recent Indigenous mass protests, such as those in the Peruvian Puno region in 2022, highlight the ongoing exclusion of indigenous peoples in Andean countries. Moreover, ethnic and class prejudices, discrimination, and "interactional inequalities" (Araujo 2009; **Frei and Orchard** 2023) have persisted, impacting lower social strata broadly and Indigenous populations specifically. Research has identified several factors contributing to this persistence, including the "colonial mindsets" of elites, power struggles over control of resource-rich areas, electoral politics, and inadequate planning and implementation by regional governments.

Andean countries continue to exhibit significant territorial inequalities. The most prominent example of this disparity is the enduring contrast in human development between prosperous urban centers and the poorer rural hinterlands, where a substantial portion of the indigenous population resides (ECLAC 2019). Substantial disparities also persist in terms of access to public services among different subnational entities within these countries (Modrego and Berdegué 2015), particularly in the healthcare sector (**Chazuffi and Leyton** 2023). Service provisions depend on the strengths of regional coalitions vis a vis the central government, electoral calculus, and the regions' economic value, among others. Fiscal decentralization reforms introduced in Andean countries to ensure a more equitable distribution of revenues from the extractive sector (such as the mining tax ("Canon") in Peru and the direct tax for hydrocarbons ("Impuesto Directo a los Hidrocarburos"/IDH) in Bolivia), have instead widened territorial gaps (**Jiménez** 2021). Within urban areas, social inequalities are intensifying, with marginalized areas affected by illegal activities such as drug trafficking and illegal land trafficking, alongside increasing individualization at the expense of community organization and social cohesion (**Torres et al.** 2021).

Gender inequality has recently deteriorated across the Andean region (ECLAC and UNICEF 2020). Women face pervasive discriminatory social and political practices, including femicide—violent killings targeting women because of their gender. Typically responsible for reproductive roles, women also bear a disproportionate impact from climate change and increased burdens due to the economic crises, amongst others (**Maldonado and Schorr** 2023). New regulation to protect women has been adopted, but remains mostly unimplemented as political symbols or as "aspirational laws" (Htun and Jensenius 2020).

Resource economies as the Andean ones are susceptible to economic, social, and political challenges, often referred to as the "resource curse" (**Damonte and Schorr** 2021). One aspect of this "curse" is economic volatility stemming from fluctuating global demand patterns: Boom periods are inevitably followed by downturns. For Andean countries, the boom phase concluded around 2013, leaving significant gaps in their fiscal accounts. Populism, authoritarianism, and corruption among rent-seeking politicians are other manifestations of the resource curse in the region. The end of the economic boom has also underscored deficiencies in the regional state's redistributive capacities. Fiscal systems continue to be regressive, placing disproportionate burdens on poorer segments of the population, thus perpetuating wealth concentration among the affluent and in certain territories (Valdés Valencia 2017). Governments, even those professing to be "progressive", have failed to address this issue, often due to the influence of powerful economic elites (Crabtree et al. 2023). Economic elites have not only shaped tax policies to maintain or increase their privileges, gains, and profits. They also influenced legislation concerning mining (Durand 2016), energy (Bril-Mascarenhas and Madariaga 2017), and water regulation (Damonte and Oré 2014), cementing and enhancing social asymmetries in the region to the disadvantage of the poor and less powerful.

The external shock of the COVID-19 pandemic has further worsened economic conditions in the region. In 2020 alone, the average gross domestic product (GDP) in Latin America plummeted by 7.7%, marking the largest decline in 120 years. Consequently, poverty and income inequality in the region have, again, markedly increased: Poverty rose from 30.5% in 2019 to 33% in 2020 (extreme poverty increased from 11.4 to 13.8%), and the GINI index rose by an average of 0.7% (ECLAC 2022). Furthermore, the COVID-19 pandemic revealed dramatic disparities in access to effective healthcare and educational opportunities. Prolonged school closures and challenges in accessing internet-based education exacerbated educational inequalities, significantly hindering the educational prospects of an entire generation (ECLAC and UNESCO 2020). Concurrently, while significant portions of the population face deterioration, wealth concentration in Latin America has intensified during the pandemic (Bull and Robles Rivera 2020; Cárdenas et al. 2020).

Extractivism had particularly adverse effects on local populations directly affected by extractive industries. These include environmental problems arising from using toxic substances in extraction, leading to public health challenges, especially in areas with a long history of extraction (e.g. in La Oroya and Espinar in Peru; **López** 2023).

Environmental and health challenges are also prevalent in regions where small-scale gold mining has surged, particularly in the Amazon region in Bolivia, Colombia, Ecuador, and Peru (**Damonte and Schorr** 2022). Other negative impacts include increased deforestation of the Amazon rainforest and a critical loss of biodiversity, as well as the rise in criminal violence, human trafficking, forced prostitution, and widespread human rights violations. Regional states have, so far, been unable or unwilling to address these problems, primarily due to the increasing political power of gold miners (**Damonte** 2018).

In "extractive territories," local communities compete with governments and national and transnational companies over natural resources. Extraction is frequently imposed on these communities by the state, ignoring or manipulating participatory processes, despite Andean states having formally adopted international regulations on consultation processes with indigenous peoples as outlined in International Labour Organization Convention No. 169 (**Flemmer** 2022). As a result, local conflicts around extractive projects have multiplied during the latest boom period. In Peru, social contention over mining has become the leading type of conflict (**Paredes** 2022).

Empirical research indicates that only a minority of these conflicts are about environmental justice or completely opposing extractive activities (**Orihuela et al.** 2022; Arellano-Yanguas 2011). Generally, local communities engage in contentious negotiations with the state and companies over redistributive issues and how they can materially benefit from the projects, reflecting a general tension between economic and ecological concerns in Andean countries and territories. However, these conflicts and negotiations occur within profound power imbalances, usually to the disadvantage of local peasants. Protests often result in some form of dialogue (in Peru, "mesas de diálogo") that rarely provide long-term solutions to the conflict (**Preciado** 2023). Commitments generally remain unfulfilled, often leading to further conflict and enhancing distrust among local people in a region characterized by extraordinarily low levels of public and interpersonal trust (Ciudadanía 2019).

During the recent boom period in the Andes, companies have increasingly engaged with communities near their projects, primarily to prevent or mitigate social resistance (Frederiksen 2018). Companies often assume state responsibilities when attempting to secure a "local license" through official Corporate Social Responsibility (CSR) activities, including investments in education, health sectors, or productive programs. They have also been found to create opaque clientelistic networks to buy off strategic local actors and reduce resistance to their investments (e.g., Cerro Verde in Arequipa; **Vela** 2021) or engage in outright illegal practices such as blacklisting and violent threats.

In sum, extractive industries have profoundly impacted community life in the Andes. They have sparked intense and sometimes violent social mobilizations, created internal divisions between supporters and opponents of the projects, and triggered processes of indigenization to strengthen local claims. Additionally, companies have increased their influence in communities where they operate, along with other external actors, such as national and international NGOs, which promote networking among affected communities and support local struggles. Some communities have begun engaging in mineral or gold mining themselves, perpetuating environmental and health problems and potentially exacerbating inequalities and conflicts among them (Ramírez 2023).

In contrast to the tensions in "extractive territories," and sometimes in response to them, governments have adopted inclusive institutional changes during the recent economic boom with the potential to reduce inequalities (**Damonte and Schorr** 2021; Kapiszewski et al. 2021). These changes included new redistributive and social policies, participatory institutions, enhanced political transparency rules, and innovative environmental legislation and territorial planning instruments. However, many new regulations have failed to achieve their intended transformative goals. The reasons for these failures are varied, including institutional designs that reinforce power asymmetries rather than reduce them (**Geng** 2024; **Preciado** 2023), and a lack of political support for adequately implementing inclusive regulations, often due to economic considerations or the influence of powerful economic lobbies. Additionally, other well-intentioned reforms, such as creating protected areas, sustainable forest management, and climate change governance, have often failed to consider local conditions and have reinforced social inequalities rather than reducing them (**Paredes and Kaulard** 2022; **Martínez** 2021).

This overview of the evolution of social inequalities in the Andean region during the recent economic cycle is not exhaustive. It only provides a glimpse into the many challenges social inequalities pose to sustainable development in the region and the various ways they evolve and transform. While the challenges remain enormous and have partly increased due to economic and ecological crises, some advancements have been made, particularly in formal institutional development. Also, some communities and regions have achieved greater economic benefits and social inclusion, shedding light on the factors promoting these positive developments, such as social capital (**Cancino** 2022), political will, and "synergistic infrastructures" connecting people and territories (**Kaulard** 2021). In a general perspective, the recent history of social inequality in the Andean region offers important lessons for achieving inclusive, sustainable development in the context of persistent inequalities. Current issues, such as the escalating climate crisis and the mandate for global energy transitions, which once again positions Latin America as a raw material provider for the world, should be assessed in light of these findings. Further research is needed to address the previously identified deep-rooted inequalities within their historical, social, economic, and political contexts, focusing on sustainability.

This section offered an overview on the trAndeS research agenda and the principal research outcomes. In the following section we address the program itself, established to promote research, training, networking and capacity building. We first explore the lessons learned regarding the establishment and management of the program. Then we proceed to discuss major achievements and challenges.

3 Lessons Learned: The Establishment and Implementation of TrAndeS

This section shares some lessons learned in establishing and managing an international academic program spanning various fields like trAndeS. In particular, we discuss several fundamental factors that helped us establish and implement our program: personal and institutional trust, shared research interests, proactive support by host institutions, and ongoing, transparent communication. We also stress the need for an adequate partner, enough time for the project's preparation, and a "trial and error" approach open to continuous evaluation and eventual adjustments.

Trust is probably the most crucial ingredient in any form of cooperation, typically gained through repeated interactions on a level playing field. In the case of trAndeS, we were fortunate to build on a long history of collaboration between social scientists at PUCP and the Institute for Latin American Studies at FU Berlin. The experience of working together successfully over the years and across different projects significantly facilitated the new collaboration that would become trAndeS. The founding members of trAndeS were confident that the collaboration would lead to significant results in training and research and that the partnership would progress smoothly without major obstacles. Moreover, the trust within the team greatly facilitated exchange and coordination during the project's preparation phase despite spanning different geographic areas and time zones and occasionally requiring timely and flexible responses, even beyond official working hours.

The Peruvian-German project team not only drew on a trusted relationship. They also shared research interests and foci. Previous collaborations between PUCP and FU Berlin focused consistently on issues of social inequality, providing a strong foundation for establishing trAndeS. This shared focus facilitated the development of the program's research agenda, curriculum, and the selection of scholarship projects in a relatively short period of time. Having a common conceptual and theoretical language is invaluable in interdisciplinary projects, fostering clarity and efficiency—a crucial factor given the time constraints typical of such collaborations.

In addition to personal trust and shared thematic foci, the previous cooperation had fostered "institutional trust" in the host institutions, PUCP and FU Berlin. Besides motivated and knowledgeable individual researchers, international academic cooperation also requires supportive and capable host institutions. Universities involved in such projects must manage complex administrative tasks such as international money transfers, scholarship programs, providing infrastructure, hosting visiting researchers, and producing accountability reports for donors, among other responsibilities. FU Berlin and PUCP are top-tier institutions in their countries, renowned for their professional administrations and extensive experience in international networking and project management. Knowing that the hosts would have the capacity to support and execute an international project significantly facilitated the project preparation phase and its subsequent implementation.

Choosing the Peruvian PUCP as a partner for international academic cooperation focused on social inequalities and sustainable development was also based on other compelling reasons. PUCP stands out as Peru's top university and ranks among the best in Latin America. Rooted in Jesuit traditions and committed to liberation theology,[2] PUCP operates as a nonprofit private institution, contrasting with the profit-driven model prevalent in the region's private education sector. This commitment to education as a public good is exceptional in Peru and the Andean region. Moreover, PUCP plays a pivotal role in shaping Peru's future leaders, positioning our program to have a tangible impact through training these individuals. The university also has an extensive regional network, including the leadership of the Red Peruana de Universidades (RPU), an alliance of the most important public and private universities in the country committed to improving higher education and research. With these attributes, PUCP was the ideal partner, serving as a key hub for achieving our research and program objectives in the Andean region.

Trusting relationships, including institutional trust, capable host institutions, and a shared academic framework, formed a strong foundation for the initiation of the program trAndeS. Additionally, trAndeS benefited significantly from proactive support provided by the host institutions in setting up, operating, and expanding the program. The German and Peruvian hosts contributed essential resources such as administrative capacity, infrastructure (including space and technology), and partial funding for project personnel. A pivotal contribution came from the Postgraduate School of the Department of Social Sciences at PUCP in Lima, the program's institutional anchor in Peru. They agreed to waive tuition fees for trAndeS students, allowing them to retain their full monthly stipends and reducing overall program costs.

Communication and negotiation were crucial in securing this support, as it did not materialize automatically but required ongoing dialogue between the binational trAndeS project team and university authorities. Maintaining regular communication, providing detailed information, direct engagement, and proposing specific forms of support were essential to successfully establishing and sustaining the project. Furthermore, since trAndeS operated at the intersection of various academic bodies, levels, and programs, ongoing and transparent communication was also essential to securing the support of the directors of different disciplinary Master's and Doctoral programs. These programs needed to accept trAndeS as a complementary initiative and validate its courses for credit. Once the program was operational, maintaining ongoing contact with these programs became crucial for monitoring student progress and coordinating joint activities.

For effective operations and high-quality academic training, this level of coordination and communication within the broader university community should be integrated into the preparation phase of international university collaborations and their implementation. This discussion on coordination and communication also highlights the importance of allocating sufficient time to prepare the operational aspects

[2] This term refers to a movement within Christian theology that emphasizes the liberation of oppressed people, particularly the poor, from social, political, and economic injustices. It emerged in the 1960s and 1970s, primarily in Latin America, in response to widespread poverty, inequality, and human rights abuses.

of the project, transforming the initial proposal into a functional structure capable of achieving its objectives. Establishing trAndeS required an extensive preparatory agenda, including acquiring infrastructure, recruiting staff, designing institutional and individual work packages, and establishing procedural guidelines. These guidelines had to align with the rules of partner universities and other involved organizations, especially funders, whose rules sometimes conflicted with those of host institutions or countries. For example, while hotels in Peru require prepayment, this was prohibited by German administrative regulations. An agreement was negotiated to resolve this, allowing payment upon arrival, with the Peruvian host guaranteeing payment to the hotel. Another challenge involved the stringent receipt-based accounting requirements imposed by DAAD in a country where economic formality is nearly 80%. To address this, after some coordination, we introduced supplementary receipts that were declared and signed by the user.

Project implementation also needed flexibility for adjustments based on trial and error. Some planned activities could not be implemented or modified due to unforeseen constraints. For instance, we originally planned to introduce a joint international course between FU Berlin and PUCP. However, this plan was abandoned after intense discussions due to differing academic calendars, time zones, and credit systems. We chose more flexible joint online student seminars instead.

In conclusion, the trAndeS experience highlights several factors crucial to successfully establishing and implementing an international program. These include mutual personal and institutional trust, well-matched institutions capable of handling complex tasks under tight deadlines and willing to provide material support, and open communication within the team and the broader university community. Negotiating different rule sets and adopting a flexible 'trial and error' approach were key to the project's successful implementation. Having outlined the factors that contributed to building and implementing trAndeS, we now discuss the program's achievements and challenges across its main areas of activity.

4 Achievements and Challenges

Since its establishment in 2016, trAndeS has carried out a wide range of academic and non-academic activities, and many individuals and organizations have been involved in the program. In this section, we explore the principal achievements and challenges in the program's four main areas of activity: scholarships, interdisciplinary research and training, networking, and institutionalization. For each area, we first present the main achievements and then discuss the major challenges, along with how we addressed them. Given the scale and duration of the program and the space limitations of this chapter, this is necessarily a selective overview.

Training and the Importance of "In Situ" Scholarships

trAndeS offers a variety of instruments to provide access to higher education and to support young researchers in their academic careers. In particular, the program runs a scholarship program for master's and doctoral students and a fellowship program for postdoctoral researchers from the Andean region. Between 2017 and 2026, 24 master's students and 26 doctoral students completed the trAndeS curriculum and graduated with the support of our program. 17 Master's students and 17 doctoral students were granted full-time scholarships from the program; 16 students (seven master's and nine doctoral) participated as associated students, a status that allowed them to access selective support such as funds for international conference participation or support for research and publications. Between 2017 and 2026, 33 postdoctoral researchers received a grant for a visiting stay at trAndeS-PUCP in Lima. In terms of graduation outcomes, the program was highly successful, achieving a graduation rate that surpassed that of any other program at PUCP.

In internal evaluations, the master's and doctoral students of the trAndeS program greatly valued the opportunity to study full-time at an excellent university in their own country. This combination of full-time study, quality education, and domestic access is exceptional for Peru and likely many other countries in the Global South. Peru's higher education sector is notably weak, with only a few private and public universities offering quality education. Enrollment is expensive, and there are very few private or public scholarships. The Peruvian authorities have made minimal efforts to invest in research and innovation, and many families cannot afford the tuition fees at private universities. As a result, students often choose lower-quality public universities or work full-time while studying to afford enrollment, which can compromise their academic performance. Alternatively, some students leave the country, often applying for scholarships to US universities. However, not everyone can study abroad, especially (female) students with families who find it difficult to relocate.

The in situ scholarships awarded by trAndeS provided these students the opportunity to complete their master's or doctoral studies at home while staying with their families. TrAndeS students have consistently emphasized that the opportunity to receive exceptional doctoral training within the country, without needing to go abroad, is the most important feature of the trAndeS training program. In situ scholarships also have a structural effect: they help reduce academic brain drain, a significant issue in countries with weak higher education systems like Peru and other Andean nations. The combination of an excellent university and a in situ scholarship program proved highly effective for academic cooperation aimed at improving higher education, promoting sustainable development and reducing social inequality.

The trAndeS Postdoctoral Fellowship Program offers two-month "writing fellowships" at PUCP to postdoctoral researchers from universities and other research institutions in the Andean region. These fellowships allow researchers to focus entirely on their academic publications and careers while receiving critical guidance from trAndeS senior researchers at PUCP. Collectively, supporting postdoctoral researchers in their scientific production enhances the academic output in and from the region. This support is especially vital given the current underrepresentation of scholars from Andean countries in international academic discussions. Moreover, the fellowships address the historically low levels of academic exchange in the region by integrating emerging scholars into an international academic community engaged in research and dialogue on issues of regional importance. The postdoctoral fellowship program has laid the groundwork for new collaborations within the region, resulting in several new publication and research initiatives.

According to the program's tracer studies, trAndeS grantees have been successful in their careers, typically continuing in academia or taking positions in state agencies. Both career paths align with the program's long-term objectives: trAndeS research agendas are perpetuated in regional universities, and as civil servants, alumni apply their knowledge of the relationship between social inequalities and sustainable development to improve public policies.

The trAndeS scholarship program faced several challenges. A key issue was balancing the requirement for academic excellence, as donors and host universities demanded, with the UN 2030 Agenda's mandate to 'leave no one behind.' While committed to the UN 2030 Agenda, trAndeS encountered practical limitations common to many Global South countries. Higher education and individual programs cannot fully compensate for deficiencies in primary and secondary education. Students from poor neighborhoods and marginalized provinces with inadequate primary and secondary education often struggle to access top universities. Consequently, trAndeS scholarship recipients generally came from relatively good universities and were not among the most disadvantaged in the country. Similarly, while efforts were made to encourage women's participation and offer postdoctoral fellowships to regional junior scholars, applicants still had to meet criteria of academic excellence. If they did not, fully adhering to the 'leave no one behind' principle in higher education proved challenging.

The program addressed these difficulties with a specific strategy. The selection process considered socio-economic, gender, ethnic, and regional factors to ensure representation of the Andean region and balanced gender distribution. This resulted in a positive gender balance in all master's and PhD cohorts, with some cohorts having more female students than male. Additionally, the number of scholarship recipients graduating from public universities outside Lima increased yearly. The program also considered 'academic potential' as a criterion for admission, in addition to academic performance and grades. Academic potential was assessed based on the relevance of research topics and questions to the trAndeS research agenda, as well as candidates' performance during personal interviews. Those demonstrating significant academic potential with important and underexplored research topics were admitted and received specialized training from trAndeS senior researchers to further enhance their academic performance. Moreover, the trAndeS network strategy aimed to address academic disparities by providing capacity building and facilitating exchanges for resource-constrained universities in the region (see next section).

Despite these efforts, resolving the tension between academic excellence and support for vulnerable groups remains challenging. Individual institutions and programs alone cannot fully address this issue. Donors should not expect international academic cooperation to serve as a substitute for addressing these structural deficiencies that require broader and more fundamental actions, particularly from host countries, to ensure equitable support and opportunities for all researchers.

Other challenges included the impact of the Covid-19 pandemic and delays in students' completion of their studies. Peru, along with other Andean countries, was severely affected by the virus, experiencing one of the highest death tolls globally. In Peru, educational institutions, including higher education facilities, were closed for two years. During this period, many students and colleagues faced personal losses and difficult situations that negatively impacted their studies and research. Field trips were prohibited during the lockdown, and limited internet access hindered online training. In response, the program transitioned to virtual training and established a special fund for Peruvian students to provide necessary equipment and improved internet connectivity. Additionally, students were granted extra time to complete their projects. For students facing delays due to other reasons, members of the Executive Committee actively reached out to understand their circumstances and explore ways to offer assistance according to individual needs.

Enhancing Cognitive Abilities Through Interdisciplinarity

Given the limited consideration of social and political factors in debates on sustainable development and sustainability transitions (Schorr et al. 2021), trAndeS aimed to contribute from an interdisciplinary social science perspective. Consequently, the program's teaching and research agenda was strictly interdisciplinary to achieve a comprehensive perspective on the impact of social inequalities on the opportunities of sustainable development.

Interdisciplinarity was evident in the master's and doctoral programs at PUCP involved in trAndeS, including anthropology, economics, political science, and sociology, as well as the interdisciplinary master's programs in "Management of Hydric Resources" and "Environmental Development". Students enrolled in these programs were eligible for the trAndeS curriculum and scholarships, ensuring interdisciplinarity at the core of each student cohort.

The program's teaching activities also reflected this interdisciplinary approach. In the two compulsory trAndeS courses, experts from various disciplines addressed sustainability and sustainable development issues from their unique disciplinary perspectives, focusing on the causes and effects of social inequalities. While it is crucial for students to gain a strong foundation in a specific discipline, providing them with a solid command of its concepts and methods, the interdisciplinary approach ensured additional benefits. It required students to engage with different disciplinary languages and perspectives on development problems. This approach better equipped them to understand the multidimensional nature of sustainable development challenges and taught them that they can be interpreted and approached in multiple ways.

Guided by the interdisciplinary trAndeS project team, which included experts in sociology, economics, political science, anthropology, and physical geography, the MA and doctoral students were expected to apply the interdisciplinary insights gained in the classroom to their research. This approach resulted in a diverse range of research that highlights the complexity of multifactorial and multi-level issues of inequality and sustainability. The list of research topics (cf. Sect. 2 of this chapter) includes unequal access to water resources in the context of mining projects and the determinants of conflict over access to livelihood resources; gender inequalities affecting indigenous women and young female prison inmates; the opportunity for the adoption and effective implementation of new environmental regulations and new redistributive and regulatory institutions, the relationship between economic and institutional volatility, spatial inequalities within Lima, the production of social consent for environmentally harmful activities (such as large-scale monocultures), asymmetries affecting Amazonian indigenous groups, the undesirable effects of large infrastructure projects on local (indigenous) populations, etc. In addition to these individual studies, trAndeS researchers have collectively published several books, both monographs and edited volumes, addressing institution-building in the context of Andean resource economies (Damonte and Schorr 2021), the microsocial impact of inequalities (Maldonado and Schorr 2023), an overview of inequality dynamics in the Andean region (Henríquez and Vila 2023), Urban Ecology in Lima (Sabogal 2021) and Ecosystem Adaption in the Andean-Amazonian region (Sabogal 2023). The program has also published teaching and outreach materials such as course readers, working papers and policy briefs. Most of these resources can be accessed freely on the programs' internet pages.[3]

While the interdisciplinary social science perspective applied in trAndeS generated a rich body of knowledge, interdisciplinarity posed challenges. First, interdisciplinarity can be arduous and time-consuming. It means acquiring basic language skills and understanding another discipline's basic concepts and methods; therefore, researchers must leave their disciplinary comfort zone. Not all researchers are open to this. To encourage interdisciplinary encounters, trAndeS researchers have been asked to integrate reflections on underlying disciplinary assumptions into their presentations and to explain discipline-specific concepts. Being open about disciplinary specificities helped others to follow and understand and contributed to producing a comprehensive perspective on the research themes in question.

Another challenge concerned 'integral interdisciplinarity' with the natural sciences. As a program grounded in the social sciences but concerned, amongst others, with issues of the 'materiality of nature', we were keen to engage with the natural sciences and those disciplines traditionally prominent in the discussion of sustainable development and sustainability (i.e. engineering). However, exchanges with the natural sciences have been limited to sporadic meetings at workshops and conferences. The SDG Graduate Schools Alliance, the alliance of all seven SDG Graduate Schools funded by the DAAD/BMZ, was the most successful format for integral interdisciplinarity. The alliance's events facilitated collaboration between researchers from both the social sciences and natural sciences, including fields such as chemistry, pharmacology, and agricultural sciences. This collaboration enabled a comprehensive exchange on critical issues such as climate change, agriculture, plastic pollution, and waste management. Additionally, members of the Alliance participated in cross-teaching activities. For example, trAndeS delivered a lecture on sustainable development at the ROHAN summer school, which was primarily

[3] All materials and publications can be accessed on our website: https://www.programa-trandes.net/index.html.

attended by chemists. Conversely, a chemist from ROHAN provided a lecture on scientific integrity during a virtual meeting organized for the network of Peruvian universities.

The lack of integral interdisciplinarity within trAndeS did not stem from a lack of willingness on either side. Rather, it emerged as a side-effect of the different questions and methods used by the two groups of disciplines. While interdisciplinary teams should be composed according to the project's needs as defined by the research topic, for future collaboration, we need to reflect on how to benefit from an integral interdisciplinary exchange between the natural and social sciences. Especially for research committed to sustainable development and sustainability transitions, integral interdisciplinarity will be key in the future.

The Individual and Structural Benefits of Networking

Networking has been one of the core activities of trAndeS and has led to the creation of the Red trAndeS, an alliance of researchers mainly from Andean countries dedicated to issues of social inequalities and sustainable development. Established at the beginning of the program in 2016, the trAndeS network currently comprises 143 researchers from 11 countries: Argentina, Bolivia, Colombia, Chile, Ecuador, Germany, Italy, Peru, Spain, United States of America and Venezuela. In the years since then, network members have engaged in various activities such as joint publications (Velásquez-Castellanos et al. 2018; Damonte and Schorr 2021; Maldonado and Schorr 2023), panels at international conferences (e.g. at the Latin American Studies Association (LASA) yearly conference), academic workshops in Lima and Berlin, and new research projects on the Latin American environmental politics, environmental concern in Andean countries, global energy transitions in Latin America and institutional traps. Networking has produced important individual and structural effects.

At the individual level, networking has promoted careers, especially of young researchers, by providing opportunities for exchange and future collaboration. In a survey of network members, including grantees, almost all respondents rated enlarging their regional and global networks as one of the most important outcomes of their contact with trAndeS. Many also stated that the interdisciplinary trAndeS network enabled them to expand their scientific repertoires regarding topics, concepts, theories, and methods.

At a structural level, networking created a research community in a region urgently needing greater academic integration. Cooperation between universities and research centers in the Andean countries is strikingly low, as is comparative research within each country. The Red trAndeS provided opportunities for researchers from the Andean countries to gather and exchange. They were integrated into the program in different roles, as scholarship holders (MA and PhD students), postdoctoral (visiting) researchers, lecturers in our courses, and active participants in our academic events. In particular, the exchanges allowed comparative visions and sharing of knowledge on similar sustainability challenges in the different countries of the Andean region (such as the socio-ecological impacts of mining or food crops recently integrated into global commodity chains, such as quinoa).

The trAndeS network also provided the context for supporting structurally weaker public universities in Peru and other Andean countries. trAndeS researchers from PUCP and FU Berlin have offered capacity building and thematic courses in several Andean (Bolivia, Chile, Ecuador, Colombia) or other Peruvian (San Martín, Cusco, Ayacucho) universities. Within Peru, we organized workshops and courses and granted targeted trainee scholarships for professors and students from regional universities, mainly from the PUCP-led Red Peruana de Universidades (RPU). These trainees discussed their research in colloquia and workshops, participated in seminars and attended lectures. Within the Andean region, in Bolivia, Ecuador, and Chile, we collaborate with public universities to realize doctoral schools for local students and participants from other Andean countries. Besides the individual consequences, these activities have an important multiplier effect: Participants can replicate what they learned during these events in their home institutions, ultimately strengthening local research and training capacities across Andean countries.

Through networking, local knowledge produced in smaller Andean universities could be fed into global sustainability debates. Bringing local knowledge into international debates is particularly important for producing knowledge to foster transitions to sustainability. In our interconnected world, we need to examine the consequences of our behavior wherever they occur. Take, for instance, policy initiatives to promote sustainable energy transitions, as for example the transition from fossil-fueled energy to renewable energy (so-called "energía verde" (green energy)). The global race for lithium, a strategic metal required for global energy transitions, especially for batteries, has intensified lithium mining in South America, often with disastrous consequences for local communities, including the destruction of nature and wildlife, pollution of water and soil at extractive sites as well as the dispossession and usurpation of forests and territories. It is here where local inequalities, for example in land possession and land use rights at cost of indigenous communities, are entangled with global dynamics and power asymmetries. Networking contributes to making these local impacts visible and revaluing local knowledge. Doing so helps bridge the fragmented global sustainability knowledge chains, as we call them elsewhere (Schorr et al. 2021).

Setting up the Red trAndeS was not difficult due to the many previous contacts with researchers in Andean countries. Translating networking into actual knowledge production was more complex for several reasons: First and foremost, networking within trAndeS occurred under inequality and diversity conditions. As mentioned above, the higher education sector in the region has several significant limitations (except for Chile). Only a few universities have a solid research profile. In addition, public universities generally have few resources and the academic level of staff tends to be low. In Peru, for instance, 50% of all university professors do not have a master´s degree, and only 7% of all professors in the country hold a PhD. Especially universities in the provinces generally focus on local problems and examine them in individual case studies. Comparative research allowing generalizations is scarce. Moreover, these studies are often not written according to the conventions, methodological considerations, and/or format of the dominant academic journals published in English. Standard peer-review processes are lacking. As a result, networking has occurred between researchers from very different institutional and academic backgrounds and cultures, making certain types of academic cooperation challenging.

Another limitation of the trAndeS network was the lack of sufficient funding to promote research, exchange, and publications within the network. Consequently, the network relied on the contributions of its individual members to carry out activities and produce knowledge. Generally low-cost activities were organized, such as virtual workshops, or we created synergies with the activities of trAndeS at the PUCP. Researchers from the trAndeS network contributed their own resources, sourced from other funding and support mechanisms. In this low-cost way, based on individual initiatives, the network has accomplished a lot and will continue to function.

Leaving a Legacy: The Institutionalization of the Program

The SDG Graduate School funding line instructed projects to institutionalize some core elements, following a common practice in international project funding. It is a particularly complex task as institutionalization requires the host university authorities to agree to long-term structural changes and investments.

Institutionalizing the main features of trAndeS was preceded by several years of negotiation and coordination with the authorities of the host institutions and began at the very beginning of the project. At both universities, institutionalization was facilitated because the project fitted into existing strategic visions. FU Berlin and PUCP had a strong institutional commitment to the internationalization of research and education before the program began. trAndeS not only connected with these strategies but also managed to develop them further. In doing so, at both universities, trAndeS gradually became a model for international academic cooperation that combines higher education with training and research on development-related issues.

At PUCP, trAndeS acted as a sounding board to test certain internationalization tools and to establish their value as a permanent feature. In concrete terms, trAndeS supported the internationalization of training and research activities by advising on developing a long-term internationalization strategy inspired by FU Berlins' "international network university" approach. The trAndeS postdoctoral program was the first at PUCP and set the stage for creating a university-wide postdoctoral program. In addition to the postdoctoral fellowships, the international events organized by trAndeS, the research stays of PUCP researchers at FU Berlin, and the research and writing fellowships have also enhanced internationalization of PUCP.

trAndeS has also highlighted the importance and value of interdisciplinary doctoral training on sustainable development issues. The interdisciplinary research and training approach of trAndeS ties in with an ongoing strategy at PUCP to improve doctoral training and strengthen the universities' research capacity. Consequently, the Postgraduate School of the Faculty of Social Sciences did not hesitate to institutionalize the two trAndeS courses on the relationship between social inequalities and sustainable development that fit into this strategy. Most importantly, the university leadership decided to institutionalize interdisciplinary doctoral training by creating an interdisciplinary doctoral school, including a scholarship program and other instruments to promote research. trAndeS not only served as a model for this effort dedicated to societally relevant research but was also integrated into the new school in 2023. When trAndeS funding ends in 2025, the new platform will continue as a legacy of trAndeS by offering scholarships (as long as funds are available) and promoting interdisciplinary research relevant to sustainable development.

trAndeS also leaves a legacy at the FU Berlin, particularly at the Institute for Latin American Studies (LAI). The implementation of the program expanded the research and course offerings at the LAI in two respects: it added a focus on sustainable development and a focus on the Andean region. At the university level, the program contributed to the incipient debates on the global responsibility of higher education institutions. Since trAndeS made convincing progress, FU Berlin also agreed to support future involvement in international projects relevant to sustainable development by institutionalizing a half-time academic position responsible for promoting this type of international academic development cooperation.

The main challenge in institutionalizing temporary project features is generating commitment and securing long-term funding from host universities. Given the general reluctance of universities to engage in such activities, often due to a lack of financial resources, this is not an easy task. In our experience, achieving institutionalization required skillful negotiation over an extended period of time. As argued above, this commitment is easier to achieve if the project fits into wider institutional strategies and contributes to their development. On the other hand, internationalization tends to expand out of itself. Once initiated, opportunities for new projects and further exchanges and funding will usually arise. This can help to perpetuate features of temporary projects.

Another challenge in institutionalizing the project is that it lies beyond the control of those in charge of the project. Institutionalization depends not only on the strategic imperatives of the host universities but also on the support of individual authorities and researchers. It is also influenced by the availability of resources and wider political and institutional factors in the host countries. A reorientation of strategic plans by the university or new authorities may force changes in institutionalization plans. Political and economic crises, common in Andean countries, may hamper efforts to institutionalize international academic cooperation. These risks must be incorporated into any cooperative effort through thorough risk assessment and effective mitigation strategies. During our collaboration on trAndeS, for example, Peru experienced several political crises. However, the program remained unaffected due to the unwavering commitment and robust support for institutionalizing its core features at PUCP and FU Berlin. Building on this legacy, the cooperation between the PUCP and the FU Berlin is well set to continue.

5 Conclusion

This chapter presents and analyzes the achievements and challenges of the international academic collaboration "trAndeS—Advanced Studies on Social Inequalities and Sustainable Development," jointly implemented by Freie Universität Berlin,

Germany, and Pontificia Universidad Católica del Perú. The trAndeS research agenda explores how multidimensional and interdependent inequalities, rooted in historical and contemporary contexts, influence sustainable development in the Andean region. It seeks to identify the actors and mechanisms that perpetuate these disparities while addressing challenges related to economic cycles, environmental issues, and social injustices. Over the past decades, social inequalities have evolved, particularly during the global commodity boom that benefited the region. Nonetheless, research in our program shows that significant challenges persist, and new ones—such as the region's role in the global energy transition—are emerging.

In addition to promoting its research agenda, the program aims to enhance research and teaching capacities at partner institutions within the Andean region. To achieve these objectives, trAndeS employs a three-pillar approach. The first pillar focuses on postgraduate education at the Master's and doctoral levels, offering a scholarship program and a specialized curriculum. The second pillar centers on interdisciplinary research, including thematic working groups addressing key issues. The third pillar emphasizes networking, aimed at strengthening regional and institutional connections.

In this chapter, we have examined several factors that we believe have contributed to the successful establishment and implementation of the program. These include the existence of interpersonal and institutional trust, alignment in research interests, proactive support from host institutions, and consistent, transparent communication. Additionally, we emphasize the importance of identifying suitable partners, allowing sufficient time for project preparation, and adopting a "trial and error" approach that embraces continuous evaluation and adjustment.

The trAndeS program has significantly contributed to advancing higher education in the Andean region across various dimensions. By offering targeted scholarships and fellowships for master's, doctoral, and postdoctoral researchers, the program has provided high-quality education and research opportunities within the region. This initiative has addressed educational inequalities, improved academic output, and promoted regional integration while balancing academic excellence and support for underserved populations. Furthermore, through its interdisciplinary research and teaching agenda, trAndeS has generated comprehensive insights into the intersections of sustainability and social inequalities. The program has also facilitated regional academic integration and individual career development through the red trAndeS network. Notably, trAndeS leaves a lasting legacy by institutionalizing key features aligned with the German and Peruvian host universities' internationalization strategies. This alignment has led to the creation of a new postdoctoral program and an interdisciplinary doctoral school at Pontificia Universidad Católica del Perú and the expansion of sustainable development research at Freie Universität Berlin.

The SDG Graduate Schools funding line has facilitated a unique integration of research, networking, and training, fostering capacity building, quality education, and internationalization at the universities involved in the program. In the case of trAndeS, the program combined individualized training at a high-caliber university, the PUCP—particularly benefiting students with familial responsibilities—with capacity building, specifically in training doctoral candidates and young scholars in Peru and other Andean countries. Additionally, the program advanced research on a socially relevant theme, highlighting the critical role of higher education in international development cooperation.

Traditionally, universities have not been central actors in development cooperation. However, the SDG Graduate Schools funding line has positioned them at the forefront of sustainable development initiatives. The various projects under this funding line demonstrate that universities are well-suited for this role. Universities, often stable institutions in regions facing instability and in need of sustainable development, contribute through international cooperation by simultaneously impacting individuals and structures across diverse locations. Many institutions, such as Pontificia Universidad Católica del Perú (PUCP), explicitly pursue missions to advance sustainable development through education and socially impactful research. Moreover, universities serve as incubators of both material and ideational innovation. They engage with young people, including future decision-makers, and shape mindsets that promote social equality and sustainability. Furthermore, universities naturally favor international cooperation and participation in global networks. In international academic partnerships, they function as knowledge brokers and disseminators of information. As global discussions on sustainability transitions and sustainable development evolve, universities will be crucial in generating and sharing innovative multi- and interdisciplinary knowledge. In sum, universities are well-equipped to lead international sustainable development-oriented initiatives and, hopefully, will be able to continue in this type of cooperation in the future.

References

Amarante V, Brun M, Rossel C (2019) Poverty and inequality in Latin America's research agenda: a bibliometric review. Dev Policy Rev 38(4):465–482. https://doi.org/10.1111/dpr.12429

Araujo K (2009) Habitar lo social: usos y abusos en la vida cotidiana en Chile. LOM, Santiago

Arellano-Yanguas J (2011) Aggravating the resource curse: decentralisation, mining and conflict in Peru. J Dev Stud 47(4):617–638. https://doi.org/10.1080/00220381003706478

Atria J, Groll CL, Valdés Valencia MF (eds) (2018) Rethinking taxation in Latin America: reform and challenges in times of uncertainty. Palgrave Macmillan, Cham

Baquero-Melo J (2014) Layered inequalities. Land grabbing, collective land rights, and afro-descendant resistance in Colombia. LIT Verlag, Zürich

Bebbington A (ed) (2011) Social conflict, economic development and the extractive industry: evidence from South America. Routledge ISS studies in rural livelihoods. Taylor and Francis, London

Blofield M (ed) (2015) The great gap: inequality and the politics of redistribution in Latin America. Penn State University Press, University Park, PA

Boatcă M (2015) Global inequalities beyond occidentalism. Ashgate Publishing Limited, Farnham

Braig M, Costa S, Göbel B (2015) Desigualdades sociales e interdependencias globales en América Latina: Una valoración provisional. Rev Mex Cienc Polít Soc 60(223):209–236. https://doi.org/10.1016/S0185-1918(15)72136-7

Bril-Mascarenhas T, Madariaga A (2017) Business power and the minimal state: the defeat of industrial policy in Chile. J Dev Stud 55(6):1047–1066. https://doi.org/10.1080/00220388.2017.1417587

Bull B, Robles Rivera F (2020) El COVID-19, las élites y el futuro de la economía política de la reducción de la desigualdad en América Latina. Rev ECLAC 132:79–94
Cancino N (2022) ¿De qué manera el capital social aporta a la conservación de recursos? Asociatividad indígena y no indígena y adopción de prácticas sostenibles de suelo de productores de cacao amazónico peruano. Masters Thesis, Pontificia Universidad Católica del Perú
Cameron M, Hershberg E (eds) (2010) Latin America's left turns: politics, policies, and trajectories of change. Lynne Rienner, Boulder
Cárdenas J, Robles-Rivera F, Martínez-Vallejo D (2020) Élites empresariales y desigualdad en tiempos de pandemia en América Latina. Rev Española Sociol 29(3):715–726. https://doi.org/10.22325/fes/res.2020.45
Castillo M, Maldonado C (eds) (2015) Desigualdades: Tolerancia, legitimación, y conflicto en las sociedades latinoamericanas. Editorial RIL-Universidad Diego Portales, Santiago
Cazzuffi C, Leyton C (2023) Desigualdades espaciales, capacidades y desarrollo en América Latina. In: Maldonado C, Schorr B (eds) La desigualdad en nuestras vidas: Una mirada microsocial desde América Latina. Iberoamericana Vervuert, Madrid/Frankfurt
ECLAC (2019) Panorama Social de América Latina 2018. Comisión Económica para América Latina y el Caribe, Santiago
ECLAC (2022) Panorama social de América Latina 2021
ECLAC, UNESCO (2020) La educación en tiempos de la pandemia de COVID-19. Informe COVID-19 ECLAC/UNESCO. https://repositorio.ECLAC.org/bitstream/handle/11362/45904/1/S2000510_es.pdf
ECLAC, UNICEF (2020) Violence against children and adolescents in the time of COVID-19 (Covid-19 Report). https://www.ECLAC.org/en/publications/46486-violence-against-children-and-adolescents-time-covid-19
Ciudadanía (2019) Informe Nacional de la Encuesta Mundial de Valores en Bolivia. Ciudadanía
Costa S (2011) Researching entangled inequalities in Latin America: the role of historical, social and transregional interdependencies. Desigualdades Working Paper (9). Freie Universität Berlin/desiguALdades.net, Berlin
Crabtree J, Durand F, Wolff J (2023) Business power and the state in the Central Andes: Bolivia, Ecuador and Peru in comparison. University of Pittsburgh Press, Pittsburgh
Damonte GH (2018) Mining formalization at the margins of the state. Small-scale miners and state governance in the Peruvian Amazon. Dev Change 49(5):1314–1335. https://doi.org/10.1111/dech.12414
Damonte G, Orè M (eds) (2014) ¿Escasez de agua? Retos para la gestión de la cuenca del río Ica. Pontificia Universidad Católica del Perú, Lima
Damonte G, Schorr B (eds) (2021) Andean states and the resource curse: institutional change in extractive economies. Routledge, Abingdon/New York
Damonte G, Schorr B (2022) The hybrid global gold regime: a perspective from the Peruvian ASGM sector. In: Brombacher D, Maihold G, Müller M, Vorrath J (eds) Geopolitics of the Illicit Linking the Global South and Europe. Weltwirtschaft und international Zusammenarbeit, vol 25. Nomos, Baden-Baden, pp 361–384
Dietz K (2017) Researching inequalities from a socio-ecological perspective. In: Jelin E, Motta R, Costa S (eds) Global entangled inequalities: conceptual debates and evidence from Latin America. Taylor and Francis, Abingdon/New York, pp 76–92
Durand F (2013) Los Romero. Fe, fama y fortuna. DESCO: Ediciones EL VIRREY, Lima
Durand F (2016) Cuando el poder extractivo captura el Estado. Lobbies, puertas giratorias y paquetazo ambiental en Perú. OXFAM, Lima
Flemmer R (2022) Prior consultation to halt the resource curse?: potentials and pitfalls of a participatory innovation in Peru and its implications for the Andean countries. In: Damonte G, Schorr B (eds) Andean states and the resource curse: institutional change in extractive economies. Routledge, London
Frederiksen T (2018) Corporate social responsibility, risk and development in the mining industry. Resour Policy 59:495–505. https://doi.org/10.1016/j.resourpol.2018.09.004

Frei R, Orchard M (2023) La dignidad como lente analítico en el estudio de la desigualdad: reflexiones desde el caso chileno. In: Maldonado C, Schorr B (eds) La desigualdad en nuestras vidas: Una mirada microsocial desde América Latina. Iberoamericana Vervuert, Madrid/Frankfurt

Freisstein K, Mahlert B (2016) The potential for tackling inequality in the sustainable development goals. Third World Q 37(12):2139–2155. https://doi.org/10.1080/01436597.2016.1166945

Fritz B, Lavinas L (eds) (2015) A moment of equality for Latin America? Challenges for redistribution. Ashgate Publishing Limited, Farnham

Fukuda-Parr S (2016) From the millennium development goals to the sustainable development goals: shifts in purpose, concept, and politics of global goal setting for development. Gend Dev 24(1):43–52. https://doi.org/10.1080/13552074.2016.1145895

Gasparini L, Cruces G (2013) Poverty and Inequality in Latin America: a story of two decades. J Int Aff 66(2):51–63. http://www.jstor.org/stable/24388285

Geng D (2024) Tecnopolítica de la desigualdad en el acceso al agua en una zona minera: Territorios de escasez y participación tecnificada en Candarave (Tacna, Perú). Dissertation, Pontificia Universidad Católica del Perú

Göbel B, Góngora ME, Ulloa Cubillos A (eds) (2014) Desigualdades socioambientales en América Latina. Universidad Nacional de Colombia; Ibero-Amerikanisches Institut, Bogotá/Berlin

Góngora M (2015) Globalización de la naturaleza y fragmentación del derecho internacional. In: Henríquez N, Braig M, Göbel B, and Damonte G (eds) Desigualdades en un mundo globalizado. Pontificia Universidad Católica del Perú, Lima

Góngora Mera M, Vera Santos R, Costa S (2019) Entre el Atlántico y el Pacífico Negro. Afrodescendencia y regímenes de desigualdad en Sudamérica. Iberoamericana Vervuert, Madrid/ Frankfurt

Gootenberg P, Reygadas L (eds) (2010) Indelible inequalities in Latin America: insights from history, politics, and culture. Duke University Press, Durham, NC

Hall GH, Patrinos HA (eds) (2012) Indigenous peoples, poverty, and development. Cambridge University Press, Cambridge

Henríquez N, Vila G (eds) (2023) Pensando la región Andina: diálogos interdisciplinarios sobre desigualdades y desarrollo. Pontificia Universidad Católica del Perú, Facultad de Ciencias Sociales, Lima/Berlin

Heredia M (2023) Note from the editor: inequality and wealth in Latin America. Econ Sociol 25(1):1–4

Htun M, Jensenius FR (2020) Aspirational laws as weak institutions: legislation to combat violence against women in Mexico. In: Brinks DM, Levitsky S, Murillo, MV (eds) The politics of institutional weakness in Latin America. Cambridge University Press, Cambridge

Jelin E, Motta R, Costa S (eds) (2017) Global entangled inequalities: conceptual debates and evidence from Latin America. Taylor and Francis, Abingdon/New York

Jiménez JP (ed) (2015) Desigualdad, concentración del ingreso y tributación sobre las altas rentas en América Latina. ECLAC, Santiago

Jiménez JP (2021) Fiscal reforms and institutional changes in the Andean region: revenues volatility and unequal distribution of regional income. In: Damonte G, Schorr B (eds) Andean states and the resource curse: institutional change in extractive economies. Routledge, Abingdon/ New York

Kapiszewski D, Levitsky S, Yashar DJ (eds) (2021) The inclusionary turn in Latin American democracies. Cambridge University Press, Cambridge

Kaulard A (2021) La construcción de redes e infraestructuras sinérgicas del cacao en San Martín (2007–2018) Dissertation, Pontificia Universidad Católica del Perú

Kreckel R (2004) Politische Soziologie der sozialen Ungleichheit, 3rd edn. Campus Verlag, Frankfurt

López AP (2023) Habitar cuerpos y territorios enfermos: Políticas sociotécnicas frente a la problemática de salud ambiental en Espinar. Masters Thesis, Pontificia Universidad Católica del Perú

López-Calva LF, Lustig N (eds) (2010) Declining inequality in Latin America: a decade of progress? Brookings Institution Press, Baltimore
Maldonado C, Schorr B (eds) (2023) La desigualdad en nuestras vidas: Una mirada microsocial desde América Latina. Iberoamericana Vervuert, Madrid/Frankfurt
Martínez AG (2021) Relaciones sociales de poder y desarrollo territorial en la creación de áreas naturales protegidas: caso del Santuario Nacional Los Manglares de Tumbes (SNLMT). Dissertation, Pontificia Universidad Católica del Perú
McKay BM, Alonso-Fradejas A, Ezquerro-Cañete A (eds) (2021) Agrarian extractivism in Latin America. Routledge, Abingdon/New York
Mills CW (1956) The power elite. Oxford University Press, Oxford
Melamed C (2012) Putting Inequality in the post-2015 picture. Overseas Development Institute (ODI), London
Modrego F, Berdegué JA (2015) A large-scale mapping of territorial development dynamics in Latin America. World Dev 73:11–31. https://doi.org/10.1016/j.worlddev.2014.12.015
Monsalve Zanatti M (2014) Grupos económicos y mediana empresa familiar en América Latina. Universidad del Pacífico, Lima
Nash JC (1979) We eat the mines and the mines eat us: dependency and exploitation in Bolivian tin mines. Columbia University Press, New York
Orihuela JC, Pérez Cavero C, Contreras C (2022) Extractivism of the poor: natural resource commodification and its discontents. Extract Ind Soc 9:100986
Paredes M (2022) Toxic mobilization: mining, pollution and power in the highlands of Peru. Environ Sociol 9(2):136–147. https://doi.org/10.1080/23251042.2022.2124621
Paredes M, Kaulard A (2022) Forest as 'nature' or forest as territory? Knowledge, power, and climate change conservation in the Peruvian Amazon. J Peasant Stud 50(6):2210–2231. https://doi.org/10.1080/03066150.2022.2134010
Piketty T (2014) Capital in the twenty-first century. The Belknap Press of Harvard University Press, Cambridge/London
Preciado R (2023) El teatro de la negociación en la mesa de diálogo de la provincia de Caylloma-Arequipa para resolver los conflictos por los proyectos de desarrollo implementados por el Estado. Dissertation, Pontificia Universidad Católica del Perú
Ramírez T (2023) Compensando el tiempo perdido: las trayectorias adaptivas de tres comunidades peruanas afectadas por la minería. In: Maldonado C, Schorr B (eds) La desigualdad en nuestras vidas: Una mirada microsocial desde América Latina. Iberoamericana Vervuert, Madrid/Frankfurt
Reygadas L (2008) La apropiación: Destejiendo las redes de la desigualdad. Editorial Anthropos, Iztapalapa
Reygadas L (2015) Más allá de la legitimación: Cinco procesos simbólicos en la construcción de la igualdad y la desigualdad. In: Castillo M, Maldonado C (eds) Desigualdades: Tolerancia, legitimación, y conflicto en las sociedades latinoamericanas. Editorial RIL-Universidad Diego Portales, Santiago
Ross Schneider B (2008) Economic liberalization and corporate governance: the resilience of business groups in Latin America. Comp Polit 40(4):379–397. https://doi.org/10.5129/001041508X12911362383237
Sabogal A (2021) Urban ecology: a case study of Lima City, Perú. Springer, Cham
Sabogal A (2023) Ecosystem and species adaptations in the Andean-Amazonian region: traditional land-use systems in Peru. Springer, Cham
Sánchez-Ancochea D (2021) The costs of inequality in Latin America: lessons and warnings for the rest of the world. I.B. Tauris, London/New York
Schilling-Vacaflor A, Flemming R (2015) Conflict transformation through prior consultation? Lessons from Peru. J Lat Am Stud 47(4):811–839. https://doi.org/10.1017/S0022216X15000826
Schorr B (2018) How social inequalities affect sustainable development: five causal mechanisms underlying the nexus (trAndeS Working Paper Series No. 1). Lateinamerika-Institut, Freie

Universität Berlin, Berlin. https://www.programa-trandes.net/Ressources/working_papers/WP-1-Schorr-Online.pdf
Schorr B (2023) Los Andes: Cumbres de desigualdad, brechas de bienestar y sustentabilidad. In: Henríquez N, Vila G (eds) Pensando la región Andina: diálogos interdisciplinarios sobre desigualdades y desarrollo. Pontificia Universidad Católica del Perú, Facultad de Ciencias Sociales, Lima/Berlin
Schorr B, Braig M, Fritz B, Schütt B (2021) The global knowledge value chain on sustainability: addressing fragmentations through international academic partnerships. Sustainability 13(17):9930–9950. https://doi.org/10.3390/su13179930
Stampini M, Robles M, Sáenz M, Ibarrarán P, Medellín N (2016) Poverty, vulnerability and the middle class in Latin America. Latin Am Econ Rev 25(4). https://doi.org/10.18235/0012281
Stewart F (ed) (2008) Horizontal inequalities and conflict: understanding group violence in multiethnic societies. Palgrave Macmillan, London
Svampa M (2015) Commodities consensus: neoextractivism and enclosure of the commons in Latin America. South Atlantic Q 114(1):65–82. https://doi.org/10.1215/00382876-2831290
Therborn G (ed) (2006) Inequalities of the world. Verso Publishing, New York
Therborn G (2013) The killing fields of inequality. Polity Press, Cambridge
Tilly C (1998) Durable inequality. University of California Press, Berkeley
Torres D, Perleche D, Aiquipa A (2021) La producción del espacio urbano en Lima Metropolitana y el Callao: entre las informalidades y la regulación (1961–2020). Universidad Nacional de Ingeniería, Lima
Valdés Valencia MF (2017) Reducing Inequality in Latin America: the role of tax policy, 1st edn. Routledge, London/New York
Van Cott DL (2000) The friendly liquidation of the past. The politics of diversity in Latin America. University of Pittsburgh Press, Pittsburgh
Vela E (2021) La red de intercambio para el acceso al agua, alrededor del Proyecto Minero Cerro Verde en la Sub Cuenca del Río Chili en Arequipa, Perú. Masters Thesis, Pontificia Universidad Católica del Perú
Velásquez-Castellanos IO, Schorr B, Damonte GH (eds) (2018) Growth, inequality and the challenge for sustainability: in a post-boom scenario in the Andean Region. Fundación Konrad Adenauer/trAndeS, La Paz
World Commission on Environment and Development (1987) Our common future. Oxford University Press, Oxford

Bettina Schorr holds a Ph.D. in Political Science from the Universität zu Köln, Germany. She is a lecturer at the Institute for Latin American Studies at Freie Universität Berlin, Germany, and the program director of trAndeS—Advanced Studies on Inequalities and Sustainable Development. Her work focuses on institution-building in extractive economies, environmental policies and socio-environmental conflicts in the Andean region. She is the author of various research articles and books, including "Andean States and the Resource Curse: Institution Building in Extractive Economies" (2021, with G. Damonte) and "Environmental vs. Economic Concerns in Bolivia: Does Support for Extractivism Affect Environmental Concern? (2024, with C. Marull)

Lea Caanitz holds a Master's degree in Interdisciplinary Latin American Studies with a focus on transformation and development, and a combined Bachelor's degree in Spanish Philology, Latin American Studies, and Communication and Publication Science. Her master's thesis examined care migration and entangled inequalities between Latin America and Germany. Since 2023, she has been working as a research assistant at trAndeS, the Postgraduate Program on Sustainable Development and Social Inequalities in the Andean Region. Her academic interests include migration, sustainable development, social policy, gender, and structural inequalities.

The Doctoral Studies Support Programme on Environmental Peace and Development in Colombia (DSSP)

Dennis Avilés-Irahola, Laura Calderón, Tomás León Sicard, Christian Petersheim, Rosario Rojas-Robles, Emilia Fernengel, Carolina Tobón Ramírez, and Eva Youkhana

Abstract The Doctoral Studies Support Programme on Environmental Peace and Development in Colombia (DSSP) is a bilateral school led by the Institute of Environmental Studies (IDEA) at the National University of Colombia (UNAL) and the Center for Development Research (ZEF) at the University of Bonn (UoB). It seeks to integrate inter- and transdisciplinary approaches in examining the connections between environment, territory, development, and conflict, while aligning research with the Sustainable Development Goals (SDGs). The program also focuses on fostering international academic collaboration, critically reflecting on higher education's role in advancing the SDGs, and enhancing information systems and data management capabilities. Through their shared commitment to critical research, alternative development models, and environmental thinking, IDEA and ZEF have gained valuable insights throughout the program's implementation and evaluation. This process has also facilitated the building of strategic alliances and the launch of innovative initiatives. The implementation of the DSSP also presented challenges arising from differing approaches and perspectives. Key lessons learned include the importance of cross-disciplinary collaboration, participatory research, and integrating local knowledge to address environmental issues. Ultimately, the DSSP highlights mutual enrichment and the pursuit of environmental peace in Colombia by fostering harmonious relationships between people and between them and nature goods.

D. Avilés-Irahola (✉) · C. Petersheim · C. T. Ramírez · E. Youkhana
Center for Development Research (ZEF), University of Bonn, Bonn, Germany
e-mail: davilesi@uni-bonn.de

L. Calderón · T. L. Sicard · R. Rojas-Robles
Institute for Environmental Studies (IDEA), National University of Colombia, Bogotá, Colombia

E. Fernengel
TRA Present Pasts (TRA5), University of Bonn, Bonn, Germany

C. Brandt et al. (eds.), *Implementing Sustainability and Internationalization in Higher Education*, World Sustainability Series, https://doi.org/10.1007/978-3-031-97837-1_3

1 Introduction

The collaboration between the Institute of Environmental Studies (IDEA) at the National University of Colombia (UNAL) and the Center for Development Research (ZEF) at the University of Bonn (UoB) began in 2018 as one of the seven Bilateral SDG Graduate Schools supported by the German Academic Exchange Service (DAAD). At that time, we could not have anticipated how enriching the joint program—the Doctoral Studies Support Program on Environmental Peace and Development in Colombia (DSSP)—would become. Neither the coordinators at IDEA nor at ZEF had navigated such a joint collaboration before, making this a venture into uncharted territory. By that time, however, Germany's commitment to supporting Colombia's peace efforts, including through research, was already well underway.

For example, the peace process that began in 2016 and culminated in the historic 'Final Agreement for Ending the Conflict and Building a Stable and Lasting Peace' (signed in November 2016) was consistently supported by the German-Colombian Institute for Peace (CAPAZ), which is also funded by the DAAD. Since its inception, in addition to establishing strong ties with CAPAZ, the DSSP built connections with other academic organizations involved in bilateral cooperation such as the Leibniz Centre for Agricultural Landscape Research (ZALF) and the activist group Colombia Paz (COLPAZ). These partnerships facilitated the exchange of information and the organization of joint events, including the 2022 presentation of the Truth Commission Report in Bonn. Together with CAPAZ, the Bonn International Center for Conflict Studies (BICC), and the University for Development Studies (UDS), the DSSP fostered robust collaborations aimed at advancing environmental, interdisciplinary, and transdisciplinary research on peace and conflict.

The DSSP addresses core interests shared by Colombia and Germany for decades: the sustainable use of invaluable natural goods, the strengthening of peace efforts, and the protection of the environment. This bilateral graduate school benefits from the extensive reach of UNAL, with its nine campuses across Colombia. As a result, the DSSP is able to extend its academic activities from Leticia in the Amazon region in the south to San Andrés in the Caribbean in the north, engaging students from diverse regions (see Map 1).

The DSSP's trajectory (2018–2025) has aimed not only to feed the research programmes of universities in Colombia and Germany but also the decision-making of communities and local and national authorities. To that end, it involves the study of different productive and extractive activities, the analysis of their territorial environmental effects and the recognition of different forms of knowledge, management and appropriation of the territories by populations and communities historically settled there.

The DSSP pursues the following core objectives:

- To introduce interdisciplinary, integrative, and innovative approaches, concepts, and methods in the study of the relationships between the environment, territory, development, and conflict.

- To follow a research agenda aligned with the Sustainable Development Goals (SDGs).
- To internationalize science and promote comparative analysis between regions in collaboration with transnational and regional partners.
- To explore and critically reflect on the role of higher education in advancing the SDGs.
- To consolidate and expand information systems and strengthen data management capabilities.

These objectives are primarily accomplished through the following:

- The academic program for doctoral and master students from diverse UNAL faculties. These programs are structured around a research agenda organized into four key themes: the relationship between conflict and the environment in its various forms; rural development models and their key actors; access to land and territorial rights; and local populations, knowledge, and education.
- Support for IDEA's Environmental Information System (SIAMI), which was developed with technical and intellectual assistance from ZEF, enabling it to now operate independently.
- The internationalization of teaching and research, supported through summer and traveling schools held in both Germany and Colombia, which also involved students from the University for Development Studies (UDS) in Ghana. Furthermore, the DSSP facilitated the production of publications through joint collaborations between Colombian and German partners, as well as with academics from other Latin American, African, and Asian countries.

This chapter provides a comprehensive overview of the DSSP's objectives, activities, and lessons learned. It emphasizes the importance of examining mutual learning processes, highlighting the remarkable added value of this cooperation, while also addressing the linguistic, epistemological, and administrative challenges encountered along the way.

Following this introduction, we place the DSSP program within the Colombian context, exploring the country's natural conditions and resource wealth. We trace the historical exploitation of nature since colonization and examine its impact on both human and non-human populations. This section also offers a brief chronicle of the Colombian conflict, introduces key concepts from Latin American environmental thought, and outlines the progress made in implementing the research agenda across its four thematic areas.

Next, we delve into the program's academic content, support for SIAMI, and internationalization activities. We then reflect on the lessons and challenges encountered during the program's implementation, focusing on the co-production of knowledge through inter- and transdisciplinary practices. We address the difficulties in translating theoretical and critical frameworks into practical applications and examine the dynamics and contradictions that arise in North-South relations, specifically between Germany and Colombia. The chapter concludes by discussing the practical implications and insights gained from our reflections on the complexities of implementing a Colombian-German program that explores the relationship between the environment and peace in Colombia through the lens of environmental thinking and environmentalism.

2 Environmental Peace in Colombia

The Colombian Environmental Context

Colombia, a megadiverse country and global biodiversity hotspot, features five main biogeographic regions (The Andean, Caribbean, Pacific, Colombian Amazon, and Orinoquia plains) and boasts high ecosystem diversity due to its complex geology and tropical climate, encompassing nearly all planet's biomes (Armenteras et al. 2013; Etter et al. 2006). In addition to its rich biodiversity, Colombia is home to a diverse mix of mestizo human populations, living side by side, including individuals of European descent, Indigenous peoples, Afro-descendants, native Creoles, and Raizales. Currently, the country is inhabited by 115 Indigenous groups and is home to approximately 65 living indigenous languages, although Spanish remains the predominant language (IWGIA 2024; Freixes 2024).

The history of Colombia's settlement is deeply shaped by a persistent conflict over land ownership. Following the Treaty of Tordesillas (1493–1494), signed between Spain and Portugal with papal approval, Indigenous Colombians were forcibly removed from their lands and subjected to extermination. To replace the Indigenous labour force, the Spanish relied on African slaves to work on plantations, and their descendants now comprise 10% of the nation's population. Legal instruments like the *Encomienda* system and Royal Grants were established to legitimize the violent dispossession of land. This pattern of exploitation continued throughout the colonial period (sixteenth to eighteenth centuries) and was not reversed during the Creole uprisings of the nineteenth-century independence movement.

Following the onset of the independence conflict (1810–1819), land was granted to powerful military, political, and ecclesiastical figures, systematically excluding peasants, Afro-Colombians, Creoles, and Indigenous populations. As a result, Colombia entered the twentieth century with a severe imbalance in land ownership. Large landowners relegated vast peasant populations to smallholdings, often located on steep mountain slopes or less fertile jungle land. Despite multiple attempts, Colombia was never able to implement an agrarian reform that would return land to its rightful owners. Throughout the twentieth century, the landowning elites thwarted any political efforts to address this issue. This persistent concentration of land is reflected in the country's Gini index for rural property distribution, which was recorded at nearly 0.89 in 2014 (Oxfam 2017). Despite efforts at land redistribution during the first two decades of the twenty-first century, land remains concentrated in the hands of a few (Pachón 2021).

Since the 1950s and 1960s, peasant rebellions, often in the form of guerrilla groups, became entangled with the production and trade of illicit crops such as marijuana, coca, and poppy. The conflict escalated as paramilitary groups sought territorial control in areas of economic interest, like mining and drug trade corridors, while state sectors sometimes failed to act or even facilitated these illicit activities. This contributed to a complex cycle of violence. In an attempt to resolve the conflict, a peace agreement was signed on November 24, 2016, between the Colombian government and the Revolutionary Armed Forces—People's Army (FARC-EP). However, opposition from political forces, the assassination of 355 former FARC combatants, and ongoing criminal activities by other violent groups have kept the conflict alive in some regions.

Despite decades of violence, Colombia, with nearly 50 million inhabitants, has built a resilient economy driven by exports like oil, coal, minerals, coffee, flowers, and bananas. The country's breathtaking natural beauty is also fostering the growth of its tourism and service sectors, creating new opportunities and providing sustainable alternatives to the illegal economies that once fuelled the armed conflict.

Latin American Environmental Thought

The conflicts and potential of Colombia can be understood through the lens of Latin American environmental thought, a perspective that highlights the complex relationships between ecosystems and cultures (Ángel 1993, 1995, 1996, 2000; Boff 2000; Leff 2011; Noguera 2004; Carrizosa 1996, 2001). This approach challenges both traditional development models (Escobar 1988) and European epistemological colonization (de Sousa Santos 2011). Environmental thinking offers a broader view of the Colombian conflict, questioning how humans inhabit the world and providing responses rooted in philosophy, poetry, and collective discourses centred on respect, love, and solidarity with both humans and non-humans. For instance, environmentalism encourages reconsidering the sacred plant coca from diverse perspectives, recognizing water as essential for life and advocating for its free access, valuing biodiversity in all its processes, perceiving soil as the "theatre of life" and humanity's heritage, defending free access to seeds, and rethinking the interconnectedness of life on the planet.

This is not merely a decolonization of global environmental thinking, but an effort to open the door to theoretical and practical reflection on various aspects of urban and rural life, reinterpreted through the biophysical and cultural lenses of local Indigenous, Afro-Colombian, and peasant communities. As theoretical progress is made in recognizing the importance of the symbolic world in shaping an environmental perspective, there is also growing momentum in the fight for climate justice and food sovereignty among Latin American peoples. These philosophical and ethical foundations have guided the research, teaching, and outreach efforts of the Institute of Environmental Studies (IDEA) at the National University of Colombia, since its founding in 1989, and continue to serve as the theoretical pillar supporting the academic activities of its members.

Environmental Peace: Peace with Human and Non-human Nature

Understanding the environment as an interconnected ecosystem-culture unity offers a comprehensive perspective in which non-human nature is seen as the foundation for life and societies, making it indispensable and irreplaceable. This view transcends the traditional mindset of nature as merely a resource to be exploited for profit, consumption, and economic gain. Instead, the environment is the result of the complex relationships between societies and their natural surroundings, shaped by the ways these connections are woven together.

In Colombia, colonial, capitalist, and neoliberal forces have imposed hegemonic relationships between ecosystems and society, deepening an economic model based on extractivism and unequal exchange of resources. The high consumption standards of the Global North are built on the exploitation and impoverishment of resource-rich, yet impoverished, countries like Colombia, leading to the loss of biodiversity and the livelihoods of many local communities. The economic, social, and power structures in Colombia, rooted in these injustices—including the exploitation of human and non-human nature for resources like oil, mining, and illicit crops—, have contributed to the ongoing violence and conflict. Despite resistance from governments and corporations continuing extractivist practices, Colombian society is gradually recognizing the need for reconciliation with both human and non-human nature as a crucial step toward achieving peace (Rojas-Robles 2018).

3 The DSSP'S Journey Towards Fostering Environmental Peace

The Peace Agreement presented Colombia with an opportunity to break the cycle of environmental degradation, regional inequality, and poverty by addressing territorial issues through effective state intervention and the participation of local groups (Castiblanco et al. 2021). Despite the above-described challenges, peacebuilding has continued through the active involvement of society, supported by public and private universities, former combatants, and international cooperation, including important contributions from the German government (Gómez Muñoz 2022).

In this context, the bilateral program between ZEF and IDEA has emerged as a key initiative, capitalizing on the intersection of their research areas. IDEA approaches environmental issues from an integrated perspective, while ZEF explores development concepts through a global, multidisciplinary framework, addressing topics such as land use, agriculture, climate change, water and energy management, and governance. Both organizations share critical perspectives on the environment, peacebuilding, and development in Colombia, collaborating on a joint research agenda rooted in a theoretical-practical, inter- and transdisciplinary approach.

This section provides a brief overview of the program along the following three areas: the implementation of the academic program, support for IDEA's SIAMI initiative, and the internationalization of teaching and research.

The Academic Program for Doctoral and Master Students from Diverse Faculties at UNAL

In November 2019, a participatory workshop among well-stablished academics, activists and students in Bogotá resulted in a research agenda, which has been evolving along four main themes: (i) the relationship between conflict and the environment in its various forms; (ii) rural development models and their key actors; (iii) access to land and territorial rights; and (iv) local populations, knowledge, and education. The programme has fostered doctoral and postdoctoral research and, since its second phase (2021), master's research also. In addition, eight (8) term papers have been produced by groups of doctoral students from different disciplines with the aim of encouraging inter- and transdisciplinarity addressing complex issues within the research agenda.

In order to exemplify the topics that have been put into practice in each of the research lines we present below a systematisation of the 48 projects that were developed till 2024, which correspond to master, doctoral and postdoctoral research projects of fellow students of the program. Annex 1 describes the projects in more detail.

Firstly, the relationship between conflict and the environment theme focuses on the territorial and environmental peacebuilding processes during the post-agreement period. Topics explored include the impact of war on forests and natural parks, the propaganda used by the FARC, and the role of cartography and hyperlocal journalism in peacebuilding efforts. Additionally, postdoctoral research in the Colombian Pacific region has highlighted the resilience of communities in Nuquí (Chocó) and the effects of eco-geno-ethnocidal practices against Black communities in Buenaventura, linked to road modernization and armed incursions. Other postdoctoral researchers have identified gaps in Development Plans with a Territorial Approach (PDETs) and Action Plans for Regional Transformation (PATR), incorporating perspectives of environmental justice and rights (Sánchez-Supelano 2021).

A significant body of research also addresses conflicts arising from competing interests over land and natural resources. Studies include the environmental challenges of hydrocarbon extraction in the Orinoco wetlands, deforestation in the Andes-Amazon corridor, and water theft in the Caribbean region (Cascaloa marshes). One postdoctoral study in 2021 analyzed environmental governance and the role of new community participation models in resolving conflicts, specifically in the Sierra de la Macarena.

The second line of research focuses on rural development models and the role of various actors, particularly in the context of agroecology. This work spans a wide range of scales, from national public policy proposals to local projects in communities across departments such as Nariño, Valle del Cauca, and Boyacá, as well as indigenous populations like the Nasa community. The primary analytical categories include food sovereignty, agroecological restoration, territorialization, biocultural memory, and agroecological market networks. Research projects have explored alternative rural development models, such as participatory territorial metabolism in the upper

Bogotá River basin, Amazonian tourism based on birdwatching, and a strategy for using fish co-products by women in Tumaco.

In addition to these contributions, doctoral research in engineering has applied quantitative tools to assess the impacts of climate variability, land-use changes on the hydrological cycle in Tolima, and ecological restoration in upper Andean areas. Another doctoral study, within the Public Health program, has analysed the relationship between rurality and the Colombian health system. Postdoctoral research has further expanded the understanding of local alternatives to rural development, identifying dignified coexistence forms linked to agroecological practices in Bogotá and exploring biocultural innovation in Colombian Pacific communities focused on biodiversity-based productive processes.

Urban planning and sustainability research has also been consolidated within this line, particularly through doctoral work in Bogotá. This has led to the development of a territorial evaluation and planning model based on urban metabolism and a historical-environmental analysis of air pollution-related health issues in the twentieth century. Research on green supply chains has explored their impact on the sustainable performance of Colombian manufacturing companies, and current doctoral research is investigating social representations of environmental contamination by medicines. These studies collectively contribute to a more sustainable and equitable approach to development in both rural and urban contexts.

The third line of research focuses on access to land and territorial rights, with contributions from law and political science aimed at understanding key phenomena such as land distribution and its environmental determinants, territorial configurations resulting from the expansion of the agricultural frontier, and the constituent power of 21st-century social movements. Researchers at the doctoral and master's level have examined specific scenarios, such as the Interest Zones for Rural and Economic Development (ZIDRES) and the Peasant Reserve Zones (ZRC), exploring these issues from the perspectives of legal and peasant justice.

The fourth line of research on local populations, knowledge, and education explores alternative forms of knowledge and their application in pedagogical models and didactic strategies. Specific contributions have been made to teaching processes in Bogotá and surrounding areas, focusing on environmental and counter-hegemonic education (Youkhana et al. 2018). As part of this work, a software prototype has been developed to measure the impact of research related to sustainability and the re-appropriation of knowledge.

Additionally, three doctoral studies examine the emotional and multidimensional relationships between communities and nature. The first study explores territorial expressions of mourning and related emotions, the second analyses collective environmental actions in the Iguaque Páramo from an affective perspective, and the third investigates the aesthetic material production of indigenous communities in the Colombian Orinoco and Amazon regions through art and ethnography.

The aforementioned research projects are carried out across the extensive network of UNAL branches, as illustrated in Fig. 1.

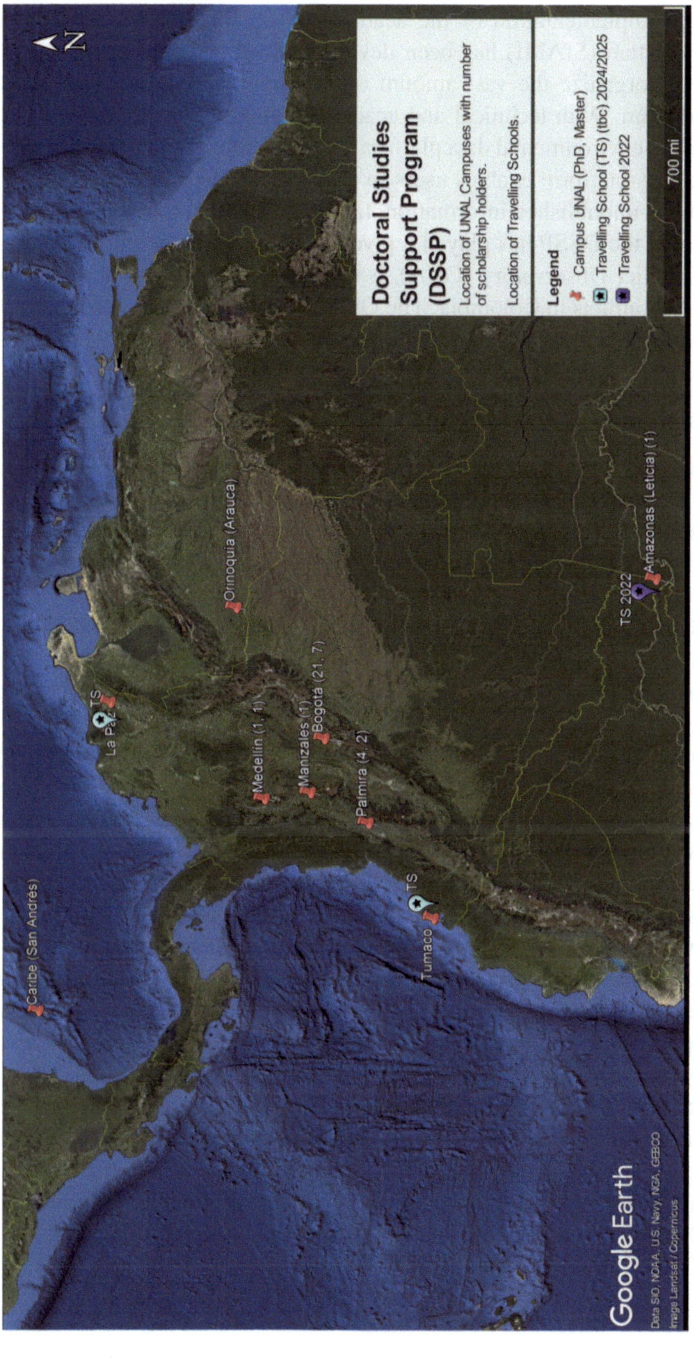

Fig. 1 Location of UNAL Branches and Traveling Schools and number of PhD and master scholarship holders

IDEA's Environmental Information System (SIAMI)

Alongside the implementation of the academic program, IDEA's Environmental Information System (SIAMI) has been developed to support an open data policy and effectively organize the vast amount of information generated by the DSSP academic program. With technical and academic support from ZEF, SIAMI functions as a virtual environmental data platform that systematizes both spatial and non-spatial data. This platform enables users to easily visualize, share, and access both documented and unpublished information, facilitating further research and analysis.

Additionally, the DSSP has played a pivotal role in strengthening IDEA's Environmental Conflict Observatory (OCA), a vital platform for raising awareness of environmental conflicts in Colombia. The OCA seeks to foster the positive transformation of these conflicts by systematizing case studies and producing comprehensive, relevant, and accessible data to inform potential solutions.

Internationalization of Teaching and Research

The most tangible outcomes of the internationalization of teaching and research include the Summer and Traveling Schools, as well as the publications produced by international researchers. The Summer Schools, held in Bonn for doctoral and master's students, combined academic activities with exchange opportunities between Colombian students and researchers in Bonn. Additionally, the Traveling Schools in Colombia featured teaching from academics from Colombia, Germany, and Bolivia. The most recent Traveling School, held at the UNAL branch in La Paz, also involved Ghanaian students from the University for Development Studies (UDS).

DSSP publications primarily feature researchers from Colombia and Germany but also include contributors from various countries across multiple regions. Notably, the proceedings of the 'International Symposium on Environmental Thinking and Alternatives to Development' (Revista UNAL 2021) showcase articles from invited guests across three continents. Most of these publications are accessible through IDEA's *Revista Gestión y Ambiente*. Similar events, though smaller in scale, have been organized and co-organized, focusing on decolonial, transdisciplinary approaches to addressing alternative socio-ecological and political models in support of environmental peace. "In summary, the inter- and transdisciplinary research agenda, along with the support for SIAMI and the internationalization of teaching and research promoted by the DSSP, tackles key aspects of Colombia's environmental challenges. These efforts yield relevant and impactful research that benefits both local communities and the global academic community. The results of the 2023 evaluation of the research agenda—published online in both Spanish and English in 2024—highlight the contributions of individual projects to the four main research themes and emphasize unresolved questions that the academic community can continue to address in support of Colombia's peace process.

4 DSSP's Learning Process and Challenges

The establishment of the DSSP, grounded in shared agreements on the importance of critical research, alternative development models, and environmental thinking outlined in the previous section, sparked a learning process that tested the very critiques of knowledge construction and research practices agreed upon by IDEA and ZEF. This section highlights how the DSSP has added value to the work of each partner—demonstrating that the whole is greater than the sum of its parts—and the challenges it has encountered, and continues to face, in aligning academic expectations between Colombia and Germany.

Learning by Doing: Co-production of Knowledge Through Inter- and Trans-disciplinary Practices

A critical examination of conventional development, driven by economic growth and the exploitation of both nature and human beings, calls for the exploration of alternative approaches. This requires transformative action that goes beyond mere explanatory or interpretive research. While there was no prior experience of international academic partners contributing to environmental peace, the DSSP partners recognized the need to examine the relationships between human and non-human nature to realize territorial rights and build socially, politically, economically, culturally, and ecologically sustainable societies.

Rather than focusing solely on human processes and the satisfaction of human needs, we embraced a broader understanding of 'sustainability'—one that prioritizes the limits of nature over the economic values assigned to its use and exploitation (Arias 2017; Chan et al. 2020; Rojas-Robles 2023). As a result, the research has emphasized the empirical, experiential, and critical study of reality across diverse disciplines, rather than adhering to the epistemological framework of economic rationalism. IDEA's experience in monitoring environmental conflicts, combined with ZEF's expertise in multi- and interdisciplinary approaches to the SDGs, led us to approach environmental peace in Colombia through inter- and transdisciplinary methods, fostering alliances with both academic and non-academic actors.

Interdisciplinarity is more than different disciplines with the same epistemological underpinnings engaging in data exchange to solve political problems or serve market trends (Mollinga 2020). It is rather about a collective mode of knowledge production which addresses theoretical and conceptual problems and engage in transformative action. On the one hand, the DSSP brings in lecturers and experienced researchers from diverse specialties such as Biology, Anthropology, Economics, Gender, and Statistics, among others. On the other hand, it designs new modules and courses based on an intensive exchange about theoretical concept and build a common language which allows to think outside the disciplinary box. This jointly created "in-discipline" enables us to leave the academic bubble, work context-driven and engage in transformative action.

The students selected as scholarship recipients come from diverse fields such as engineering, medicine, anthropology, biology, law, and economics, though most pursue doctoral programs that adopt an inter- and transdisciplinary approach. This wealth of expertise is most effectively brought together in feedback sessions on individual doctoral and master's projects. These sessions often involve discussions on research methodology, the integration of local knowledge gained during fieldwork, and the exploration of gender dimensions, as well as connections to historical and global inequalities. In this context, natural sciences and the measurement of indicators are framed within the socio-cultural and normative contexts in which they are studied. Additionally, the practice of writing term papers in interdisciplinary student groups fosters the development of professionals who are equipped to navigate the complexity of development issues beyond the confines of their own disciplines.

Transdisciplinarity is another pillar in our academic and teaching praxis. From the DSSP, we approach local knowledge based on the recognition that armed conflicts, extractive activities, mega-constructions and intensive agrochemical use have local and differentiated effects on human and non-human beings and on socio-economic inequalities, such as access to and use of land and water. These local and differentiated impacts are expressed in the knowledge of communities as historical and lived constructions, are inscribed in the bodies of men and women, and are reflected in the power relations between them and with external actors. Therefore, it makes a lot of sense to overcome the Cartesian paradigm of knowledge production, that separates mind and body, alienates our thinking from lived and felt experiences and in which the production of knowledge is reduced to reason.

The transdisciplinary approach is in line with the decolonial thinking that guides our research design and methodologies, insofar as it challenges the Eurocentric view of the world, critiquing its claims of universality and the detachment of humans from the earth, the cosmos and other living beings (Mignolo and Walsh 2018).

A significant portion of doctoral research within the DSSP is carried out in close collaboration with communities, using various participatory methodologies. This engagement is further enhanced through extended field visits to both rural and urban areas, a key component of IDEA's academic program. These long-term interactions help establish deeper connections with the communities, resulting in more meaningful and context-specific research outcomes. While having a common aim proved essential, it was not enough for co-constructing new knowledge. We learned that the fast-paced nature of academia can lead to rushed responses if we do not take the time to align on the questions. For instance, we often receive answers researchers expect to hear, rather than those that emerge from shared reflection. Our goal is to break this pattern of interaction between researchers and communities, and while we do not yet have all the answers, we understand the importance of taking the time to build knowledge collaboratively. Ultimately, inter- and transdisciplinarity is about lowering barriers to understanding, fostering new insights, actions, and emotional connections.

As examples of our exploration and implementation of inter- and transdisciplinary research toward environmental peace, we highlight two key events. The first is the DSSP-organized "International Environmental Thinking and Alternatives to Development Symposium" (November 2019), where participants from Colombia, Germany, India, Nigeria, Ecuador, Brazil, and Bolivia, representing diverse academic backgrounds and activist organizations, shared their perspectives on environmental thinking and critiques of development from both South-South and North-South viewpoints. Another significant event involved rural community representatives in the workshops of the project "Participatory Monitoring of Biodiversity Changes in Colombia: Construction of Alternatives to Environmental Conflicts." These workshops highlighted that community monitoring of conflicts must be understood through the lens of strengthening community organization (social fabric) and the defence of territorial rights. Local actors and communities are challenging conventional development and conflict studies, revealing academia as often too short-term and disconnected from addressing real-life problems.

Within the DSSP, two projects were developed in collaboration with local organizations to monitor environmental conflicts: "ColPaMon" and "Techniques and Technologies for Participatory Monitoring of Environmental Conflicts in Colombia and Germany." The former, a two-year project, sought to develop responses to environmental conflicts in Colombia, introducing tools from the OCA and SIAMI to communities affected by mining and oil exploration. The latter, a one-year study, focused on identifying techniques and technologies for monitoring environmental conflicts in both Germany and Colombia, involving students from both countries.

The failure to account for power relations in analysing environmental degradation and its unequal exploitation poses a significant risk of worsening inequalities and endangers those who put their livelihoods and lives at stake in defence of their territories. In the post-peace agreement context, where violence persists, understanding ecosystem functions and their utilization is closely tied to recognizing the local knowledge and practices embedded within specific communities' territories. Local leaders with whom we collaborated emphasized that communities' lives are deeply connected to their territories and their resistance efforts. They also acknowledged that science can contribute valuable methods and techniques for identifying evidence of environmental degradation, misuse of natural heritage, and violations of territorial rights. Furthermore, they highlighted that universities have the potential to create networks that unite diverse territorial experiences and amplify their struggles at national and international levels, supporting both territorial defence and the protection of lives threatened by such efforts in Colombia.

IDEA and ZEF have confirmed through practice that critical and transformative research on epistemic and material violence is forged through partnerships with various actors and the integration of diverse knowledge and experiences. Recognizing collective actions and responsibilities during the armed conflict, along with the emergence of new interest groups in former guerrilla-controlled areas, has underscored the importance of building strategic alliances. This collaborative approach is essential not only to fully understand the complexities of the conflict but also to contribute meaningfully to the peacebuilding process.

Critical Reflections on the Implementation of the DSSP

The DSSP's initial commitment to bridging academic research with local knowledge and perspectives—strengthened through the inter- and transdisciplinary work of grant-holders in collaboration with communities and local organizations—requires a critical examination of our own situated knowledge. This collaboration serves as a valuable mirror, allowing us to scrutinize the limitations and assumptions embedded in our perspectives (Sprague 2016) and to confront the challenges of co-constructing knowledge (Avilés Irahola et al. 2022).

Situated knowledge, shaped by the social, cultural, and political contexts, as well as the life experiences of those who express it as 'truth,' surfaced in the tensions between IDEA and ZEF in academic practice. These tensions reflect the distinct theoretical and practical approaches each institution brings to the collaboration. UNAL, and IDEA in particular, are recognized for their strong academic reputation and the social sensitivity of their faculty and students, which often translates into research 'for' social transformation. In contrast, ZEF, based in Bonn, adopts an interdisciplinary approach rooted in academic partnerships with universities in the Global South, though it is less embedded in the everyday realities of local communities, engaging with them primarily through specific research projects. Consequently, concepts like 'environmental conflict' and 'transformative research' take on different meanings in Colombia compared to Germany, shaped by their respective contexts and experiences.

In practice, IDEA students and researchers have embraced an ambitious methodological approach aimed at achieving transformative outcomes. These efforts include promoting agroecological practices within peasant communities and supporting peace initiatives led by former guerrillas now participating in the peace process. This approach, which integrates *senti-pensar* (feeling-thinking) with action, reflects a broader scientific ethos in Colombia, particularly at UNAL, where professional and personal commitments are deeply intertwined. The concept of *senti-pensar*, introduced by Orlando Fals Borda, originates from peasant fishermen on Colombia's Caribbean coast (Botero Gómez 2019). It emphasizes "thinking with the heart and feeling with the head," and, far from being an abstract concept, embodies the resilience of grassroots communities in defending their territories (ibid).

In contrast, the academic environment at ZEF in Germany places greater emphasis on conventional academic standards, such as meeting project deadlines, publishing research, and disseminating results. This often leads to a focus on analysis and findings, sometimes overlooking the emotional dimensions of the research problem. The former Department of Political and Cultural Change (ZEFa, now ZEF-CPC), where the DSSP is situated, serves as a space for discussing alternative development paradigms. Here, the balance between traditional academic formality and emerging research methodologies fosters nuanced theoretical frameworks that reflect varying degrees of emotional and intellectual engagement with the research topic.

The use of different languages—English in Germany and Spanish in Colombia—also contributes to varying interpretations of key concepts. For instance, terms like 'challenge' and 'empowerment' are understood differently depending on context. What may be considered an environmental challenge in Germany is often addressed through institutional frameworks, whereas in Colombia, it typically involves a protracted struggle for basic rights, reflecting a deeper social conflict. Similarly, 'empowerment' might be seen in the North as a concession to disadvantaged populations in the South, while in the South, it is understood as an internal process of self-affirmation, identity-building, and the exercise of territorial rights.

These differences in situated knowledge became apparent during discussions, for example, of term papers. Collaborative work among students, supervised by IDEA and ZEF researchers, not only challenged students' disciplinary boundaries but also highlighted the need to adhere to specific presentation formats, problem formulation, timelines, and rigorous research methodologies. This raised several critical questions: What is the right balance between incorporating lived experiences, the researcher's positionality, and the objective analysis of data? To what extent do the voices of local communities resonate through the researchers, and vice versa? Should term papers be treated solely as academic exercises, or should they meet certain criteria to be published and contribute to public debate? So far, these questions have been addressed on a case-by-case basis through ongoing conversations between supervisors and students.

ZEF's and IDEA's research approaches, both rooted in critical perspectives on development such as feminism and environmental thought, as well as in the equal valuation of knowledge produced in the Global South and North, recognize the epistemological tensions that arise from diverse experiences and contexts. This shared critique of Western knowledge hegemony has prompted us to avoid replicating academic curricula that are disconnected from local and regional realities. For example, in an interdisciplinary course on the history of feminism, we challenge the conventional understanding of feminism as a series of four historical waves defined by women's movements in the Global North. Instead, the course highlights *'feminismos sin ola'* (wave-less feminisms) from the South—movements characterized by territorial and communal struggles, shaped by class, ethnicity, caste, race, and the global power imbalances between the North and South.

Another example can be found in the 'Culture, Environment, and Society' modules undertaken by DSSP students at IDEA. These modules introduce foundational concepts in environmental development, focusing on human-nature interactions. By drawing from multiple disciplines—including biology, ecology, ecological economics, and anthropology—these modules avoid hierarchical distinctions between fields, using them as tools tailored to specific cultural contexts. The syllabus, including topics and suggested readings, is often adapted through critical discussions with students. The goal is not to reach predetermined conclusions but to foster a debate on how theoretical approaches can be applied to concrete, local situations.

The DSSP: A Development Cooperation Program Integrated into Bi-national Relations

In this section, we reflect on the dynamics and contradictions within North-South relations, specifically between Germany and Colombia. German development cooperation seeks to achieve the 17 Sustainable Development Goals (SDGs), primarily through the work of the Federal Ministry for Economic Cooperation and Development (BMZ), in areas such as pandemics, climate protection, displacement and migration, and digitalization. German cooperation in Colombia has spanned over fifty years and currently focuses on three main themes: peacebuilding and conflict prevention, environmental policy and the sustainable use of natural resources, and sustainable economic development.

In particular, for SDG 4 on quality education, German cooperation aims to strengthen teaching, research, and university administration at higher education institutions. The DAAD, as the largest funding organization for international academic cooperation, is also the largest provider of scholarships for international students in Germany. Colombia is the Latin American country with the second-largest number of scholarship holders in Germany, with 3510 students in 2022.

The DSSP, funded by the DAAD as part of its Bilateral SDG Graduate Schools programme, not only provides an opportunity for collaboration between German higher education institutions and academic peers in Latin America to advance the 2030 Agenda, but also serves as another link in the broader relationship between these key economic partners. However, this relationship is complicated by the growing demand for non-renewable energy resources, such as Colombian coal and oil, from several northern countries, including Germany. This demand challenges the extractivist development model that does not fully align with the rhetoric of sustainability and environmental protection.

Furthermore, while Colombia and Germany have engaged in discussions about developing alternative and renewable energy sources, the focus has often been on large hydroelectric dams, which account for 70% of Colombia's energy matrix. Despite their benefits in providing electricity to the population, large hydroelectric projects are increasingly questioned due to their significant social and ecological costs (Rojas-Robles and Santander-Durán 2021).

Once again, we are confronted with a dichotomy of visions and interests that converge on the need to support peacebuilding in Colombia but fail to reach a consensus on the "what" and the "how." Do we understand sustainability as improving conditions for resource exploitation to drive economic growth in the Global South and provide resources to the Global North, despite the high environmental and social costs? To what extent can science contribute to a transition toward development that centers on both human and non-human life, moving beyond the myth that economic growth is a prerequisite for well-being? The implementation of the DSSP has become a platform for expanding these global debates, offering examples of research focused on the sustainability of life, rather than growth at any cost.

The notion of sustainable development, framed as reconciling economic processes with the conservation of nature, has proven insufficient to address the environmental crisis that has been warned about for over thirty years (Leff et al. 2003; Riechmann 1995). In this context, alternative perspectives call for moving beyond the purely technical aspects of economic development to include the ecological, complex, systemic, cultural, diverse, and ethical dimensions. One such perspective is environmental sustainability (Rojas-Robles 2023). This prompts us to constructively question whether we can support context-driven methodological proposals like Social Life Cycle Analysis (LCAs) (Martínez-Vallejo et al. 2021) and Social Metabolism (González de Molina and Toledo 2014). These approaches integrate social, economic, environmental, and cultural variables, deconstructing the concept of "development" and fostering the construction of life itself.

The debate has become increasingly complex as political polarization in Colombia has intensified since the signing of the Peace Agreement. President Gustavo Petro's government has proposed "Total Peace" as a state policy, enabling negotiations with various armed actors within a framework of respect for human rights. However, his administration faces resistance from traditional political, economic, and military elites, including large landowners and anti-gender movements. On environmental matters, the new government has proposed strengthening civil society, particularly peasant and indigenous groups, to organize territories around water and environmental justice. While some segments of Colombian civil society support this approach, others remain sceptical of a president who was once part of a guerrilla group (*Movimiento 19 de Abril M-19*) and is challenging historical economic interests in unconventional ways.

We believe that our role in higher education remains the co-production of non-hierarchical, transdisciplinary knowledge and the construction of evidence that empowers communities to defend their territorial rights. However, we acknowledge that we are deeply embedded in a historical process of power struggles over land, water, and other natural resources, as well as over discursive definitions of "development." While this debate is global—evident, for example, in discussions on degrowth and environmental justice in Germany—in Colombia, it manifests most starkly in the decline and degradation of biodiversity, which directly impacts the quality of life for its inhabitants.

Moreover, the links between the demand for biodiversity as a "resource" from the Global North and the material violence it has generated and continues to generate in Colombia cannot be ignored. These realities compel us to critically examine the unequal exchange conditions of goods and capital at the global level, particularly the consequences of the economic development model in countries whose institutions have been weakened by decades of violence and exploitation. The exploitation of women's labour and the marginalization of ethnic groups, as well as those with little economic relevance in mining, agro-industrial, and forest exploitation processes, are tied to global dynamics that urgently require analysis and action.

5 Practical Implications and Conclusions

The DSSP addresses the complexity of the relationship between the environment and peace in Colombia from the perspective of environmental thinking and environmentalism. Drawing from individual and joint reflections, the actors involved in the DSSP recognize the need to move beyond a merely critical approach aimed at "fixing" the flaws of international development policies. Instead, they advocate for a collaborative re-imagining of these policies. In implementing the programme, it has been essential to question the concept of development that is focused on economic growth, anthropocentric and destructive views on ecosystems and local cultures. We have confirmed that contributions to environmental peace stem from embracing 'other' forms of knowledge and indigenous perspectives on issues such as agriculture, local organization, the defence of life, and the concept of *buen vivir* (good living).

A key lesson learned from the DSSP is that its critical contribution to achieving sustainable development goals, especially quality education, is most impactful when rooted in Colombia's unique realities and challenges. The 'cross-fertilization' of knowledge between IDEA and ZEF—the exchange and integration of diverse perspectives that led us to new understandings—has created tensions in the program's implementation, particularly in how issues like empowerment, poverty, and research are conceptualized. However, these tensions have proven to be valuable, offering the potential to transform our perspectives and encouraging us to view the world through the eyes of the 'other.'

Another important lesson learned has been the participatory approach in shaping the DSSP's research agenda. Involving academics, activists, and grassroots organizations in identifying research topics ensured that we addressed relevant and meaningful questions within the Colombian context. As a result, doctoral, post-doctoral, and master's research topics seamlessly aligned with the research themes, offering concrete insights into each issue. The challenge of systematizing the research has been partially addressed through the evaluation process conducted in 2023, which not only assessed the progress on specific themes but also examined the significance of the findings in light of the new direction set by the government of Gustavo Petro (2022–present). We hope the results can be applied at three levels. First, academically, by providing concrete evidence of processes that promote or hinder environmental peace, while generating new research questions. This includes challenging stagnant knowledge in isolated disciplines by fostering cross-disciplinary dialogue. Shifting focus from outcomes to processes has encouraged students to engage in a series of seminars and consultations with IDEA and ZEF professors, promoting collective research construction. The term papers also exemplify the students' growing interdisciplinary capacity. However, there is still considerable progress to be made in fully integrating inter- and transdisciplinary research practices.

The second level pertains to society: the outcomes of the research agenda can be used by actors involved in environmental conflicts, agriculture, and various sectors such as legal, educational, and technological, to drive socio-environmental transformation (Specific examples can be found in the Evaluation of the Agenda, published on the DSSP webpage). In this regard, we have built upon the foundations of IDEA's master's program, where many faculty members and students have brought their experience of working with and for communities into the DSSP. The third level focuses on decision-makers, as research findings can inform legal and project implementation processes. For instance, DSSP doctoral students have conducted evidence-based analyses to support legal actions defending territorial rights, advocate for the inclusion of environmental modules in basic education, and contribute to the implementation of PDETs and tourism-related initiatives.

The key challenge for the DSSP moving forward is ensuring the continuity of the epistemological framework it has developed and extending its impact across science, society, and policy design. This continuity must be navigated within the evolving socio-environmental dynamics that have emerged since the Peace Agreement was signed, including the emergence of new actors, changes in land use, deepening historical inequalities, and the intensification of extractivist practices. These challenges are further complicated by a new government that is striving for 'total peace' within a global economic context that continues to rely on Colombia's vast natural resources.

Another ongoing challenge we face as a bilateral program is building an academic community where all researchers—across diverse fields of knowledge and regions of Colombia—feel a sense of belonging that shapes their academic and professional careers. The opportunity to join an international network and enrich postgraduate studies with global perspectives through the platform offered by ZEF is a unique benefit for participating researchers. However, factors such as work or family commitments, often not fully supported by the scholarships, can prevent them from engaging in the program's activities, especially once their formal affiliation ends. Many scholarship recipients must maintain other sources of income during their studies. This situation calls for alternative strategies to enhance the "sense of belonging" and strengthen the community formed by the program after its completion. In the DSSP's case, such strategies have included involving researchers in other IDEA activities—such as courses, lectures, discussions, thesis supervision, or inviting them to participate as experts in IDEA radio programs based on their areas of expertise. These efforts also contribute to the program's long-term impact.

As the program enters its final year of implementation, we reflect on how academia in Germany can benefit from the knowledge generated by the DSSP and how the program can share its insights with German academia and development actors. We are also exploring how the DSSP can serve as a model for building a South-South-North cooperation program, an initiative identified as necessary in discussions with actors from other regions of the Global South, particularly in Africa.

In conclusion, we highlight the mutual enrichment between the IDEA and ZEF teams throughout the program's implementation. This collaboration has fostered the exchange of pedagogical practices and deepened theoretical debates on development and its alternatives for peace-building. Notable outcomes of this exchange include the inclusion of gender perspectives—previously absent in IDEA's teaching plan—from both feminist and gender-critical viewpoints from the South and North. Similarly, the IDEA perspective on environmental thinking, particularly the concepts of *senti-pensar* and the environment as a complex web of ecological, social, political, and cultural interactions, has influenced the academic environment at ZEF. This success has been made possible through mutual respect and a shared commitment to collective planning, research, and evaluation. Ultimately, the human dimension and the friendships that underpin these cooperation processes highlight our joint dedication to environmental peace and development, based on harmonious relationships between people and between them and the goods of nature.

Abbreviations

BICC	Bonn International Center for Conflict Studies
BMZ	Federal Ministry for Economic Cooperation and Development
CAPAZ	German-Colombian Institute for Peace
COLPAZ	Colombia Paz
DAAD	German Academic Exchange Service
DSSP	Doctoral Studies Support Program on Environmental Peace and Development in Colombia
FARC-EP	Revolutionary Armed Forces—People's Army
IDEA	Institute of Environmental Studies
SDGs	Sustainable Development Goals
SIAMI	Environmental Information System
OCA	Observatory of Environmental Conflict
UDS	University for Development Studies
UNAL	National University of Colombia
UoB	University of Bonn
ZEF	Center for Development Research

References

Ángel A (1993) La trama de la vida. Bases ecológicas del pensamiento ambiental. Series: Special documents. Ministerio de Educación Nacional Colombia, Instituto de Estudios Ambientales (IDEA), Universidad Nacional de Colombia

Ángel A (1995) La fragilidad ambiental de la cultura. Universidad Nacional de Colombia

Ángel A (1996) El reto de la vida. Una introducción al estudio del medio ambiente. Ecofondo

Ángel A (2000) La Aventura de los símbolos. Una visión ambiental de la historia del pensamiento. Series: Construyendo el futuro. 5. Ecofondo

Arias J (2017) La sostenibilidad justa como paradigma sistémico ambiental. Gestión y Ambiente 20(2):232–243. https://doi.org/10.15446/ga.v20n2.64257

Armenteras D, Cabrera E, Rodríguez N, Retana J (2013) National and regional determinants of tropical deforestation in Colombia. Reg Environ Change 13(6):1181–1193. https://doi.org/10.1007/s10113-013-0433-7

Avilés Irahola D, Mora-Motta A, Barbosa Pereira A, Bharati L, Biber-Freudenberger L, Petersheim C, Quispe-Zuniga M, Schmitt C, Youkhana E (2022) Integrating scientific and local knowledge to address environmental conflicts: the role of academia. Hum Ecol 50:911–923. https://doi.org/10.1007/s10745-022-00344-2

Boff L (2000) La dignidad de la tierra. Ecología, mundialización, espiritualidad. La emergencia de un nuevo paradigma. Editorial Trotta

Botero Gómez P (2019) Sentipensar. In: Kothari A, Salleh A, Escobar A, Demaria F, Acosta A (eds) Pluriverse: a post-development dictionary. Tulika Books and Authorsupfront, pp 302–305

Carrizosa J (1996) La evolución del debate sobre desarrollo sostenible. La Gallina de los huevos de oro: debate sobre el concepto de desarrollo sostenible 5:44–68

Carrizosa J (2001) ¿Qué es ambientalismo? La visión ambiente compleja. Revista Gestión y Ambiente 4(1):21–25. https://revistas.unal.edu.co/index.php/gestion/article/view/88811/75421

Castiblanco Rozo C, Forero C, Hernández C (2021) Retos ambientales en la construcción de paz. In: Castiblanco Rozo C (ed) Consecuencias ambientales de una paz que no llega. Instituto de Estudios Ambientales, Universidad Nacional de Colombia, pp 39–102

Chan KMA, Boyd DR, Gould RK, Jetzkowitz J, Liu J, Muraca B, Naidoo R, Olmsted P, Satterfield T, Selomane O (2020) Levers and leverage points for pathways to sustainability. People Nat 2(3):693–717. https://doi.org/10.1002/pan3.10124

de Sousa Santos B (2011) Epistemologías del Sur. Utopía y Praxis Latinoamericana 16(54):17–39

Escobar A (1988) Power and visibility. Development and the invention and management of the third world. Cult Anthropol 3(4):428–443

Etter A, McAlpine C, Wilson K, Phinn S, Possingham H (2006) Regional patterns of agricultural land use and deforestation in Colombia. Agr Ecosyst Environ 114(2–4):369–386. https://doi.org/10.1016/j.agee.2005.11.013

Freixes J (2024) Indigenous languages in Colombia. A heritage in danger. Available via Colombia One. https://colombiaone.com/2024/10/11/colombia-indigenous-languages-heritage/. Accessed 15 April 2024

Gómez Muñoz K (2022) Sustentabilidad ambiental de los proyectos productivos de los(as) excombatientes de las FARC-EP. Master Thesis, Universidad Nacional de Colombia. https://repositorio.unal.edu.co/handle/unal/81988

González de Molina M, Toledo VM (2014). The social metabolism. A socio-ecological theory of historical change. Springer

IWGIA. The International Work Group for Indigenous Affairs (2024) In: Mamo D (ed) The indigenous world 2024. Report. IWGIA

Leff E, Argueta A, Boege E, Goncalvez C (2003) Más allá del desarrollo sostenible. La construcción de una racionalidad ambiental para la sustentabilidad: una visión desde América Latina. Medio ambiente y urbanización 59(1):65–108

Leff E (2011) Economía ecológica, racionalidad ambiental y sustentabilidad. Reconstruction of the lecture given at the IV Ibero-American Congress on Development and Environment, Bogotá, Colombia, 5–9 Oct 2009. Sustentabilidad(es) 2:1–13. http://www.sustentabilidades.usach.cl/sites/sustentable/files/paginas/02-07.pdf

León-Sicard T (2010) Agroecología: Desafíos de una ciencia ambiental en construcción. In: León-Sicard T, Altieri M (eds) Vertientes del pensamiento agroecológico: fundamentos y aplicaciones. SOCLA and Universidad Nacional de Colombia, pp 53–77

Martínez-Vallejo LA, Cortés-Mora HG, Méndez-Alcázar JA, Peña-Reyes JI (2021) Un enfoque desde la sustentabilidad: análisis de ciclo de vida como herramienta para la toma de decisiones en el desarrollo de proyectos hidroeléctricos en Colombia. Gestión y Ambiente 24(2):224–237. https://doi.org/10.15446/ga.v24nSupl2.86822

Mignolo WD, Walsh CE (2018) On decoloniality: concepts. Duke University Press, Analytics, Praxis. https://doi.org/10.2307/j.ctv11g9616

Mollinga P (2020) Of binaries, boundaries and benevolence: critical interdisciplinarity in natural resources management. In: Bruns A, Kanesu R (eds) Borders in perspective, vol 5. UniGR-Center for Border Studies, pp 70–93. https://doi.org/10.25353/ubtr-xxxx-2ef3-07f3

Noguera P (2004) El reencantamiento del mundo. Universidad Nacional de Colombia

Oxfam (2017) Radiografía de la desigualdad. Lo que nos dice el último censo agropecuario sobre la distribución de la tierra en Colombia. Oxfam. https://www.oxfam.org/es/informes/radiografia-de-la-desigualdad

Pachón FA (2021) Distribución de la propiedad rural en Colombia en el siglo XXI. Revista De Econonomía e Sociologia Rural 60(4):e242402. https://doi.org/10.1590/1806-9479.2021.242402

Revista UNAL (2021) Environmental thinking, critiques on development and south-north-south cooperation proposals under construction. Gestión y Ambiente 24(1). https://doi.org/10.15446/ga.v24nsupl1.92944

Riechmann J (1995) De la economía a la ecología. In: Trotta (eds) Desarrollo sostenible: la lucha por la interpretación. Trotta, pp 11–36

Rojas-Robles R, Santander-Durán J-P (2021) Hidroelectricas, política hidroenergética y conflictos ambientales por represas en Colombia. Gestión y Ambiente 24(12):9–14. https://revistas.unal.edu.co/index.php/gestion/article/view/99655

Rojas-Robles R (2018) Ambiente y post-acuerdo en Colombia: la construcción de una paz integral y con la naturaleza no-humana. Gestión y Ambiente 21(2):183–192. https://doi.org/10.15446/ga.v21n2supl.77961.

Rojas-Robles R (2023) Perspectivas en Sustentabilidad Ambiental. Instituto de Estudios Ambientales—Universidad Nacional de Colombia

Sánchez-Supelano LF (2021) Formulación e implementación de Planes de Desarrollo con Enfoque Territorial -PDETs- desde la justicia ambiental y la teoría del goce efectivo de derechos. Gestión y Ambiente 24(2):94260. https://doi.org/10.15446/ga.v24n2.94260

Sprague J (2016) Feminist methodologies for critical researchers: bridging differences. Rowman and Littlefield

Youkhana E, Leifkes C, León-Sicard T (2018) Epistemic marginality, higher and environmental education in Colombia. Gestion y Ambiente 21(2):15–29

Dennis Avilés-Irahola holds a master's in Human Ecology from the Free University of Brussels and a Ph.D. in Development Studies from the University of Bonn. With over 20 years of experience, she has worked across agriculture, environment, political science, and gender studies, serving as a consultant and academic in various regions and countries.

Laura Calderón was a member of the DSSP coordination team at the Institute of Environmental Studies (IDEA) of the National University of Colombia from 2018 to 2024. Currently, she is a junior researcher in the Ecology and Natural Resources Management Division at ZEF, University of Bonn, working on economic incentives for conservation and the interface between biodiversity and agrobiodiversity in Colombia.

Tomás León Sicard is Agrologist, Ph.D. Professor at the Institute of Environmental Studies (IDEA) at the National University of Colombia. His areas of expertise include agroecology, agrobiodiversity, and soil science.

Christian Petersheim is anthropologist by training and Senior Researcher at the Center for Development Research (ZEF) at Bonn University. His research focuses on land use change and deforestation in southern Mexico from an ethnographical perspective. At ZEF, he is working for capacity building projects in international cooperation in higher education with partner universities in Latin America and West Africa.

Rosario Rojas-Robles is biologist, Master in Ecology, Conservation and Wildlife Management, PhD in Sciences-Biology. Associate professor at the Universidad Nacional, since 1999. Her working areas are conservation biology, ecology and environment, environmental conflicts, environmental sustainability, environmental thinking, and political ecology.

Emilia Fernengel is a cultural anthropologist working at the intersection of archaeology, heritage, and material culture. Her regional specialisation focuses on Latin America, especially Colombia, Mexico, and Peru. She is currently working at the Global Heritage Lab at the University of Bonn, a research hub for critical heritage-related and museum studies.

Carolina Tobón Ramírez has an inter- and transdisciplinary background, trained as a chemical engineer with a Master in Environment and Development and a Ph.D. in Geography. She is currently part of the coordination team of the DSSP and also a fellow of the first batch of the programme. She has focused on studies of social metabolism at regional level, sustainability, territory, integrated water management in watersheds and the role and participation of local communities.

Eva Youkhana is an anthropologist by training and professor in critical development studies. She has been working in different thematic fields, water related projects, urban and migration studies, environmental conflicts with an intersectional and material semiotic perspective. Her regional focus has mainly been on Latin America but also West Africa and Europe.

Performing Sustainability—Navigating the Complexities of a Trilateral Graduate School in Ghana, Nigeria, and Germany

Lea Frauenknecht, Martin Ringsmut, Naomi Andrew Haruna, and Sabina Appiah-Boateng

Abstract This chapter investigates the challenges and strategies of the trilateral Graduate School (GS) "Performing Sustainability. Cultures and Development in West Africa," which promotes sustainability and the Sustainable Development Goals (SDGs) across Germany, Ghana, and Nigeria. The GS's commitment to gender equality, decolonizing research practices, and fostering international collaboration is examined within the context of navigating diverse academic and cultural expectations. The article highlights the GS's efforts to integrate the SDGs into its framework, emphasizing sustainable research practices focused on community engagement and knowledge exchange. Key insights include the importance of grassroots approaches to sustainability, recognizing local interpretations of the SDGs, and ensuring research data accessibility. The GS's initiatives underscore the need for ongoing critical engagement with its theoretical perspectives on cultural sustainability, leading to a more nuanced understanding of culture in sustainable development. The findings suggest that while the GS has made significant strides in creating a supportive and sustainable academic environment, ongoing efforts are required to balance the complexities of international partnerships and a shared but critical understanding of sustainability within a postcolonial framework.

L. Frauenknecht (✉)
Institute for Cultural Policy, University of Hildesheim, Hildesheim, Germany
e-mail: frauenknecht@uni-hildesheim.de

M. Ringsmut
Department of Musicology, University of Vienna, Vienna, Austria

N. A. Haruna
Department of Industrial Studies, Faculty of Environmental Studies, University of Maiduguri, Maiduguri, Nigeria

S. Appiah-Boateng
School for Development Studies, University of Cape Coast, Cape Coast, Ghana

C. Brandt et al. (eds.), *Implementing Sustainability and Internationalization in Higher Education*, World Sustainability Series, https://doi.org/10.1007/978-3-031-97837-1_4

1 Introduction

This chapter introduces and discusses the framework of the Sustainable Development Goals (also referred to as "SDGs") Graduate School "Performing Sustainability—Cultures and Development in West Africa". We, the local coordinators of the Graduate School (also referred to as "GS") in Germany, Ghana, and Nigeria, brought together our experiences in managing the program, its students and its administration.

This chapter draws from interviews with alumni across various programs like Migration, Politics, Peace and Development Studies, Geography and Regional Planning, Music, Theatre, Museum Studies, Visual Arts, and African Studies, as well as our own experiences as coordinators in Germany, Ghana, and Nigeria. While focused on alumni perspectives, we have also included our insights to provide a fuller, more varied account of the Graduate School's impact.

We address several issues pertaining to the subject of cultural sustainability, particularly within the context of transnational research collaborations. Firstly, we discuss the process of establishing international projects, specifically the international Graduate School, which is dedicated to promoting the SDGs. In "Spaces of Exchange and Compromise: How to Build International Cooperation within the SDG Graduate School 'Performing Sustainability'?" we address issues related to exchanging and negotiating transdisciplinary and transcultural differences by highlighting the challenges and opportunities in merging diverse academic and cultural perspectives. This is followed by a section which addresses the complexities of aligning distinct university systems and the development of academic competencies. We then cover the issue of creating networks for scholarship holders and academic staff and conclude with a discussion of strategies for managing and learning from cultural differences.

Secondly, we identify the challenges encountered in such international cooperation and draw insights from the lessons learned throughout the process. "The Operative and Structural Challenges of a Trilateral Graduate School" section delves into the various issues faced by the SDG Graduate School. We begin with a discussion on gender-based challenges, including how the program addresses family-related issues and gender equality. Next, we explore efforts to decolonize research cooperation and work towards equal partnerships between institutions. We discuss the difficulties students face in starting and maintaining their thesis writing and how they navigate varying institutional requirements. We then reflect on the challenges coordinators face in managing expectations and securing additional funding, highlighting the complexity of sustaining such a trilateral collaboration as the GS "Performing Sustainability".

Lastly, we discuss the ways in which multilateral graduate schools contribute to sustainability efforts and to the achievement of the SDGs. In the section "The Promotion of Sustainability and SDGs within the GS," we demonstrate how the Graduate School integrates sustainability and the SDGs at multiple levels. It details the implementation of a dedicated data management strategy to ensure ethical and accessible data handling that counters neocolonial extractivist logics (Hansell Clark 2021). The section further describes the GS's institutional integration at partner universities and its efforts to establish sustainable frameworks beyond the initial funding phase. We discuss the GS's primary focus on SDG 16, while acknowledging intersections with other SDGs, and present four key insights gained from fieldwork, local understandings of sustainability, the processes of sustainability and the importance of community engagement. These elements underscore the GS's commitment to fostering a comprehensive and critical approach to sustainability in a globalized context.

However, before delving into the discussion of these three main questions, we provide a short overview of the context of the GS and the methodological foundation on which we developed our analysis. After this, we examine the discursive key terms that underpin the research and framework of the GS, namely, development, sustainability, culture, and performativity. To gain a deeper understanding of the interrelationships between these paradigms within the context of analyzing a transnational and interdisciplinary program, it is crucial to highlight some discursive nuances associated with these terms and how they relate to the very structures and research of the GS, before tackling the more specific questions regarding the challenges involved and lessons learned when coordinating a tri-national graduate school.

2 Context of the SDG Graduate School "Performing Sustainability"

The interdisciplinary Graduate School "Performing Sustainability: Cultures and Development in West Africa" is a collaborative training network for postgraduate students at the University of Cape Coast (Ghana), the University of Maiduguri (Nigeria) and the University of Hildesheim (Germany). The initiative is funded by the German Academic Exchange Service (DAAD). It focuses on innovative research that combines approaches from performance, arts and culture to bear on sustainable development as defined in the UN Sustainable Development Goals (SDGs). The primary goal of the GS is to foster international and trilateral academic collaboration centered on culture and sustainability. This training network is built on five key pillars: negotiating transcultural and transdisciplinary differences, bringing together three distinct academic and institutional environments, promoting capacity-building initiatives at each partner university, establishing academic and social networks, and, ultimately, providing a form of intercultural training. The ways in which these five pillars are addressed within the SDG GS will be discussed further in this article.

In order to install collaborative, cooperative and democratic mechanisms of collective decision-making, the GS comprises a board of directors, consisting of one director per university, a steering committee with three members per university, an executive board comprising four coordinators (two at the partner university in Germany and one each in Ghana and Nigeria), and two administrative positions, all of which are located at the University of Hildesheim. The current local coordinators at the University of Maiduguri and the University of Cape Coast are also alumni of the SDG GS. These different bodies are responsible for ensuring the GS's modus operandi on different levels through regular meetings. While the executive board's task is to ensure the day-to-day functioning of the Graduate School, the Board of Directors, in the presence of the four coordinators, is entrusted with important decisions with regard to the topics, activities and budget of the GS. Ultimately, the Board of Directors and the Steering Committee hold two joint biannual meetings during which they decide the long-term objectives and thematic orientation of the GS.

The GS aims to formulate international research approaches in cultural studies on the SDGs, focusing on peace and conflict studies for doctoral and master students in West Africa. The school started its operation in 2016 and will run through 2025. The first batch of students began their studies in Ghana and Nigeria in 2017. Through a structured program, 36 PhD and 30 Master students are qualified within the Graduate School's second and final phase. The University of Hildesheim (Germany) offers the option of a "cotutelle de thèse," a split-site doctoral program which provides certification for the West African PhD candidates from the two participating universities.

All students consistently mediate and disseminate the acquired knowledge throughout their future academic careers. Following the focus of the artistic research-based approach of the University of Hildesheim, program participants are encouraged to start working on projects in the educational, socio-cultural or cultural policy field to primarily implement SDG 16, but also other Sustainable Development Goals into their societies. The GS funds a yearly one-month stay for one African scholar in Germany, a yearly three-month stay of an international researcher at one of the African universities and one three-month doctoral research residency in Germany for the PhD scholarship holders. It also offers a South-South collaboration between the University of Cape Coast and the University of Maiduguri in the form of co-teaching exchanges.

Three short-term scholarships for German postgraduate students in Ghana or Nigeria were used for collaborative research projects with African students during the first funding phase. This collaboration contributed to mutual quality control of the work and links the cultures of knowledge. A unique addition made during the project's second phase is the research data management. This allows for the collation of all data from the beneficiaries onto centralized systems for future use in research projects.

Each year, the GS organizes two ten-day biannual workshops in Nigeria and Ghana, forming the central part of the training for the doctoral and master students. They comprise a theoretical examination of a subject, research methodology and advanced training in scientific research, ICT-based teaching and learning, research and data management, project management, and any identified needs for the scholarship holders. The workshops cover cultural sustainability, community development, conflict analysis, and conflict transformation. Between the workshops, a series of six online lectures take place annually. The GS also organizes monthly colloquiums followed by guided discussions. The students write assignments and receive feedback from their supervisors.

3 Development, Sustainability, Culture, and Performance—Defining and Discussing the SDG Graduate School's Operating Terms

Development

The term "development" is often used to describe the pursuit of standardized or quantifiable goals, frequently equated with notions of progress and modernization. However, this concept carries significant controversy, especially in cultural and political discourses. Critics have argued that development, rather than being a universal or neutral goal, is a social and cultural construct shaped largely by Western perspectives of socio-economic progress (Schech and Haggis 2000, xi–xii). Scholars such as Arjun Appadurai and Francis Fukuyama have challenged the top-down, Western-centric vision of development, emphasizing its varied meanings and practices as it undergoes globalization and hybridization.

Recognizing these critiques, many scholars now favor the term "transformation" over "development" to capture the complex, non-linear processes of social change. In fields like cultural studies and foreign policy, "transformation" is seen as more appropriate, as it moves away from the neocolonial and neoliberal connotations often associated with development (Heinicke 2023, 6–7). This shift also aligns with the GS viewpoint and our discussions on sustainability, where transformation emphasizes adaptive, context-specific change rather than Western-driven modernization or economic growth.

Sustainability

Sustainability refers to the practice of meeting the needs of the present generation without compromising the ability of future generations to meet their own needs (see Brundtland 1987, 3). It involves a comprehensive approach that balances social, economic, and environmental considerations to achieve long-term well-being and stability. Sustainability entails the responsible and efficient utilization of natural resources, the mitigation of detrimental impacts on the environment, the promotion of social equity and justice, and the development of resilient and inclusive communities. As asserted by Purvis, Mao, and Robinson: "The three-pillar conception of (social, economic, and environmental) sustainability, commonly represented by three intersecting circles with overall sustainability at the center, has become ubiquitous" (Purvis et al. 2019, 1). The role of culture within this model remains a subject of extensive debate. Suggestions include the addition of culture as a fourth pillar to existing models or the understanding of culture as a foundational element upon which every other pillar relies. This perspective emphasizes that culture permeates all aspects, from the social to the economic and environmental (Ododo et al. 2020). Cultural sustainability is therefore closely tied to the concept of culture.

Culture

Perhaps the most fundamental concept of the GS that distinguishes it from other SDG Graduate Schools is its focus on culture. This concept poses significant challenges in terms of definition, particularly within the trilateral context of Germany, Ghana, and Nigeria. Anthropological literature presents a multitude of definitions that attempt to encompass various aspects of culture. For example, Edward B. Tylor famously defined culture as "that complex whole which includes knowledge, belief, art, morals, law, custom, and any other capabilities and habits acquired by man as a member of society" (Tylor 1871, 1). Despite modern definitions that perceive culture as "a whole way of life" (Williams 1958, xiv), the inclination to differentiate between "high" and "low" culture continues to permeate everyday life, social institutions, and politics, particularly in the Western context. Furthermore, the tendency to equate culture solely with the arts further complicates matters. This is especially true for our cooperation partners at the University of Hildesheim, namely the Center for World Music and the Institute for Cultural Policy, whose institutional bias leans towards an understanding of culture solely as "the arts". As a Graduate School, we had to navigate these distinct and occasionally divergent understandings and will elaborate on them later. Consequently, we adopted a broad anthropological understanding of culture that explores the reasons and methods behind human actions. This approach enables us to encompass a wide range of topics, including court systems, agriculture, land development, theater, music, design, and many others. In all these instances, performance as a broader cultural concept plays a crucial role.

Performing/Performance

The concept of performance, though often associated with theater and the arts, has wider applications across disciplines. Initially tied to linguistics through J.L. Austin's idea of speech acts creating realities, performance was traditionally seen as a fleeting, present-bound act (Austin 1962, 6). Peggy Phelan reinforced this view, arguing that performance exists only in the moment and cannot be preserved without losing its essence (Phelan 1993, 146). While this transience is evident in the arts, recent perspectives expand performance to encompass everyday social life, shaping our understanding of memory, identity, and culture. Diana Taylor highlights performance as both a practice and a lens for understanding the world, making it a valuable tool for studying cultural sustainability. Within the GS, performance serves as a methodology that reflects the dynamic nature of social and cultural practices, offering a nuanced perspective on how they contribute to sustainability.

When connecting the concepts of development, sustainability, culture, and performance to the GS "Performing Sustainability," their relevance becomes evident. The broad definition of culture aligns with the diverse research projects within the GS, spanning various disciplines such as musicology, theater studies, agriculture, and political science. This definition also ties in closely with a more fluid understanding of performance as the repetition of everyday practices. Similarly, the GS promotes an understanding of development not as a merely quantifiable, economic goal, but as part of an ongoing process of negotiation.

In the following section, we address pertinent issues that arise when creating international cooperations based on these concepts, focusing on transdisciplinarity and transculturality, institutional contexts, network and capacity building, and intercultural awareness.

4 Spaces of Exchange and Compromise: How to Build an International Cooperation Within the Context of the SDG Graduate School "Performing Sustainability"?

Anyone working on a project that is not merely a one-human show is certainly aware of the fact that cooperations do not always prove to be the socially and professionally enriching, workload-sharing utopias we would like them to be, and there are various factors that can contribute to these moments of disillusion: bugs in the communication between team members, varying expectations with regards to the project outcomes, incompatible ways of doing things, and so on. This already challenging configuration of "usual" cooperation projects tends to become even more difficult when more fundamental parameters vary, such as transcultural differences, language barriers, different institutional backgrounds, and different academic disciplines. These are just some of the examples of the differences that are all combined within the cooperation of the GS, and which will be further analyzed in this chapter in order to discuss how such a trilateral cooperation can be sustainably built and continuously transformed into a long-lasting networking space.

Exchanging and Negotiating Transdisciplinary and Transcultural Differences

The GS creates an environment where academic discourses from three countries and various disciplines within the field of the humanities all come together. It can thus be described as a space within which different perspectives on both transdisciplinary and transcultural matters can co-exist, which might not necessarily be the case outside of such international cooperations, as described by one of the interviewees: "(…) SDG also helped us to make sure that we get the knowledge, and not only the knowledge, we can also have an interaction with other people whom we didn't know before." (alum 1 2022) This quote shows that the SDG GS not only provides a space for an exchange of knowledge to take place, but also enables exchanges between scholars who would otherwise never have met. For example, this is the case at the biannual doctoral workshops at the University of Cape Coast and at the University of Maiduguri Liaison in Abuja, where the scholarship holders, staff members from the three partner universities, and also international experts and institutional and political stakeholders come together to participate and exchange.

In his interview, the alumnus particularly highlights the benefits of mobility: "(…) we also benefited from visiting other countries, to see their culture, and to exchange our views, and to exchange our ideas and thoughts." (alum 1 2022) The fact that the funding of mobility between Ghana, Nigeria, Germany, and possibly other countries of origin is an integral part of the GS seems to contribute not only to academic and purely disciplinary exchanges, but also to developing a more general understanding of different worldviews and the way that people in other local contexts perceive certain issues, and to respect these differences. The interviewee describes this mutually respectful relationship between scholars on the project as a continuous example of sustainability:

> We are all graduated, but now, we have circles within. And, now, we are also changing ideas, we are learning from each other, we are sharing our thoughts with one another, which is part of sustainability. Let us continue, not stop at a certain point in time. Let the relationship continue, let the academic exchange continue, let me respect your culture and let you respect my culture. (alum 1 2022)

Within academic discourses, the GS also seems to provide the opportunity to critically discuss certain hegemonic concepts, for example the mainly Western-oriented perception of cultural sustainability itself, which can be analyzed from different national angles, as one of the alumni describes it: "(…) the way the Ghanaians see cultural sustainability is not the way Cameroonians see cultural sustainability, it is not the way Nigerians see cultural sustainability and certainly not the way Germans see it." (alum 2 2022) Although the degree to which discourses are as intrinsically linked to national identities is perhaps not as clear as the interviewee argues in her statement, the exchange of different perspectives from diverse local contexts surely feeds a broader and more accepting academic environment, as another alumnus states: "On another level you also get to appreciate the fact that we all try to learn and they have a lot to learn from each other. For instance, when we have a concept, and I talk with the others and learn." (alum 3 2022) For some, the transdisciplinary and transcultural exchange on academic topics also means evolving skills in critical thinking, while at the same time accepting ambiguities within the discursive space of the GS:

> Such a setting made me appreciate diversity a lot which is an integral part of the SDGs. As an African scholar, I became more critical about what I accept as authentic. I became especially critical about questioning how concepts and theories are explained to us and how we accept them. (alum 4 2022)

This acceptance of diversity within the project as mentioned here by one of the interviewees indeed combines two particularities of the GS. On the one hand, it is the aforementioned cooperation between three universities in three different academic and cultural settings that seems to broaden the discussion around certain discourses and concepts such as (cultural) sustainability. On the other hand, the aspect of diversity is also represented by the various disciplines of the scholarship holders and lecturers who are part of the project, as another alumnus mentions in his interview: "And then the interdisciplinary, we had a more extensive variety of disciplines of those among us who lecture us, both for the coursework and the conferences." (alum 5 2022)

While the wide spectrum of transcultural and transdisciplinary viewpoints within the Graduate School benefits an exchange of ideas as well as a critical reflection of one's own cultural and academic perspectives, it is by no means to be regarded as an easy process, as another alumna, who is now a coordinator on the project, describes:

> But then I also had to consider other aspects of academia, so other individuals, researchers, how they thought, how they think. The thought processes towards things, methodologies, were all different. What was expected in Ghana was different from what was expected in Germany and in Nigeria. So bringing that together was something totally new because we had to somehow work within the same umbrella and make sense to each other basically. That was challenging and advantageous at the same time. (alum 6 2022)

It can thus be said that, while the GS seems to provide a space for respectful and transformative transcultural and transdisciplinary exchanges, it is not to be regarded as a space which appears naturally within the realm of academic cooperation. Rather, it has to be seen as the result of a challenging collaboration between different institutions that needs to be analyzed further.

"An Oddly Unique System that Works": Bringing Together Three Different Institutional Settings

If the GS is to be regarded as a space of academic and cultural exchange between three different universities, it becomes clear that these institutions also need to move closer together in order to create this shared space. In light of the fact that universities are complex administrative apparatuses in themselves, and thus need to reframe their rigid modes of functioning if they want to collaborate, the GS's institutional setup is, of course, highly complex, challenging, and ever evolving. One of the alumni, who is now a coordinator on the project, uses a particular metaphor to describe this construct: "Three different people – I will call the universities 'people' – now. So three different people, three different ways of doing things, but trying to make it work as one person, you know. (…) one voice – but different road maps. That's a huge challenge." (alum 6 2022)

These challenges are not only perceived by the executive staff on the project, but are at the same time intrinsically linked to the scholarship holder's experiences within the project, contributing to their academic skill set:

> The major challenge was having to meet the requirements of the different universities which were a bit, and in some cases to a very large extent, different. However, the experience was rewarding in the sense that it made me work extra hard to meet the requirements of each university, making me a more resilient scholar. (alum 4, 2022)

While the cooperation between different academic institutions might come with unusual and frequent challenges which call on the flexibility of the individual universities, they also enrich both entities and individuals by making them more resilient. This is especially true regarding administrative and legal issues, such as the cotutelle contracts. These agreements allow PhD candidates—if they choose—to have their research supervised both at their home university (either the University of Maiduguri or the University of Cape Coast) and at the University of Hildesheim, ultimately receiving a PhD diploma from both institutions. While these cotutelle processes offer clear benefits, they also pose challenges for both the institutions and the candidates. PhD students must submit an additional application to the University of Hildesheim, requiring further documentation and meeting Germany's administrative requirements. Meanwhile, the legal departments of both universities, in cooperation with the GS executive team, must negotiate a joint cotutelle contract that satisfies both legal and academic standards. Throughout this process, the institutions engage in ongoing exchange and negotiation, as supervisors from Nigeria/Ghana and Germany have to align their approaches and methodologies, even within the same academic discipline, to reach a shared understanding of the candidate's research.

In this trilateral collaboration, it is crucial to consider the power dynamics and potential (neo)colonial hierarchies that may emerge in North-South academic partnerships. For example, the funding provided by German institutions can create imbalances in decision-making and influence, potentially overshadowing the priorities and methodologies of partner institutions, reflecting dynamics reminiscent of colonial relationships. Acknowledging these issues fosters a more equitable partnership that respects the autonomy and contributions of all the institutions involved. The German Academic Exchange Service (DAAD), for instance, defines the funding guidelines for the SDG GS and determines how funds can be used through the approval of the financing plan and the joint grant agreement.

Although the financial plan is developed in close cooperation with partner universities, and minor shifts between categories may be allowed, major funding guidelines are still imposed by European institutions. These institutions are often unaware of local needs and financial practices, particularly in the Global South. For instance, DAAD's invoicing requirements, such as having a tax number or responsible tax authority, can be difficult for local partners to meet as many do not have these systems in place. In such cases, it is essential to maintain an open dialogue about what is absolutely necessary and what can be adapted, while respecting the differing financial practices of the project partners.

This means that the cooperation is about creating a shared space, but also making sure that the hierarchies and power relations that made this space exist in the first place are well-balanced and actively deconstructed and power is shared. As one of the interviewed alumni puts it: "There cannot be a relationship without friction, so we need to stick on the keyword 'compromise', you need to give some room and then they give up some room, and then things will work out". (alum 5 2022) In the case of North-South cooperations, power sharing can be put into practice by various strategies: one of them is the use of funding coming from institutions in the Global North in a way that strengthens sustainable structures in the partner countries and allowing them to use it to push their own agenda.

Sustaining Knowledge and Skills in Academia: Capacity Building Within the Network

One of the main long-term goals of the SDG Graduate School "Performing Sustainability" is to create strategies for sustainable academic capacity building within the three partner universities. This objective touches upon several levels of the project. Above all, the program gives out scholarships to young researchers to enable them to pursue careers in academia. The curriculum at the home universities, the online colloquia and biannual doctoral workshops provide the PhD and MA scholarship holders with valuable thematic inputs in the areas of cultural sustainability, peace and conflict studies, cultural studies and sessions on academic skill training. The participants can then apply the contents of these workshops and teaching sessions within their own academic setting where they can also take on the role of multipliers within their own (research and teaching) community, as one of the alumni describes it: "(…) you have received a training, and the training is not for you, it's for your community, and whomever that you come in contact with." (alum 1 2022)

The fact that many GS alumni have remained in the field of academic teaching and research after completing their PhD proves that they are qualified to pursue their careers in academia. This can be shown, for instance, by referring to the example of alumnus number 1, who believes his numerous teaching activities are a direct result of his participation in the GS's scholarship program:

> I am now teaching PhD students, particularly anthropology of war and peace (…) I am also teaching (…) Nigerian cultural heritage, I am the one handling the course, because I have a PhD in cultural sustainability. All this is a result of the DAAD giving me the opportunity to be in the SDG Graduate School. (alum 1 2022)

For some, taking part in the program also represented a way of entering or re-entering academia and pursuing their career goals in this field, sometimes after leaving the university sector after their BA or MA to work in other sectors. Another alumna of the program and current project coordinator describes how she shaped her academic trajectory through being part of the GS:

> But, for me, from how they (her department, L.F.) saw me, I needed to remain in academia, but then, after my master's, I had to move from academia, get some job for myself which was non-academic. And one of the only ways to go back to academia was to go through PhD studies. (alum 7 2022)

This statement clearly demonstrates that the scholarships and staff capacities provided by the program make it possible to (re-)integrate qualified graduates and scholars into the partner universities where they can successfully continue their career.

In this sense, the program not only allows for capacity building at the master and doctoral level, but also improves staff capacities. In all three partner universities, project coordinators and—since the beginning of the second funding phase in 2021—additional RDM coordinators have been trained to accompany the organizational and RDM-related processes within the GS. Furthermore, bachelor, master and PhD students have been actively involved in the project as student assistants. Even though it benefits the project as such, capacity building seems especially crucial with regard to the transformational processes of the partner institutions beyond the program itself, since it allows for the creation of skill sets and networks that can be sustained after the end of the funding period(s). As one of the alumni states with regard to his own career as a researcher:

> (…) I came from a performing arts background and my horizon, my conceptual and theoretical, practical horizon is broad now. And together with a lot of things, it stands out. I am more successful in my career, there are a lot of extensive networks, publishing journals, getting

> research material – I wouldn't have had that network if I studied in my initial university alone. (alum 5 2022)

Another important aspect of the GS as a cooperation project is thus to create international networks for its scholarship holders and staff.

Entering and Sustaining the "Circles Within": Creating Networks for Scholarship Holders and Academic Staff

Through their respective 2- or 3-year scholarships in the GS's program, the MA and PhD students get in touch with various resource persons and their affiliated institutions both inside and outside academia. This is achieved primarily through their participation in the curriculum at the home universities, in online colloquia, the biannual doctoral workshops, conference participations, and their three-month research stay at the University of Hildesheim. Apart from the financial support and capacity building that the program provides, the opportunity to create networks is mentioned as one of the main benefits by the interviewed alumni. One alumnus, for instance, states that the "(…) Graduate School creates this very beautiful platform where people from different horizons come and meet, relationships and other alliances are created (…)" (alum 5 2022). This means that the program itself contributes to creating spaces where international networking can take place. Another interviewee describes the research stay in Germany as a key experience for his own academic networking: "In Germany, I met nine professors, and I am still communicating with them. (…) When I go through my phone, I have all their names. So, I still have contacts." (alum 1 2022)

Although these international networks are created during the time of the scholarship and the activities take place within this context, what is highlighted by most of the interviewed alumni is the way in which those networks can be sustained and benefitted from after the scholarship period. This is particularly illustrated by the already aforementioned statement by alumnus 1: "We are all graduated, but now, we have circles within." (alum 1 2022) However, it seems important to notice that these international networks can work in different ways and also fulfill various purposes. For some, having access to sustained international networks in academia can help further their career at university, as another alumna states:

> On the project I have entered networks that help my career, and this is thanks to the project. I met so many persons from different continents that are interested in my work and met so many friends that became resource persons. (alum 2 2022)

For others, these networks can help when it comes to further funding applications: "Sometimes I feel as if the fact that I got the DAAD exposure is also the reason why I'm able to win other things that I applied for (…)." (alum 8 2022)

The networks created through the participation in the GS thus seem to represent an advantage for young scholars who seek to advance in their career. However, the downside is that this clearly creates inequalities in access for those who have not had the chance to participate in such a project. Despite the fact that the funding scheme of the GS seems to sustain certain privileged and closed "circles within" (alum 1 2022), it might also benefit academic institutions and their members in a broader sense. For instance, alum 1 mentions a cooperation project between a gallery at the University of Maiduguri and the Center for World Music at the University of Hildesheim that could benefit both institutions and their affiliated researchers:

> There was a time (…) we are going to discuss and see how we are going to collaborate, between this Center for World Music and also our gallery in the University of Maiduguri. So that's how I have created a transnational network, so how they are going to assist us, or we are going to assist them, in terms of artifacts. (alum 1 2022)

An alumna and current project coordinator also mentions the creation of the GS as a network which provides exchange and training between different generations within the project and the respective institutions that they are affiliated with:

> I also foresee that, years to come, it is the same people that we are in the team with who will be heading some institutions. By that time, you have younger ones that we are training, and we would want to do some engagement, external collaborations. (alum 7 2022)

Some of the alumni also state that these international networks might not be sustained in the same way depending on their respective local and national, but also cultural contexts, as one Ghana-based alumnus points out in his interview: "With many people in Nigeria, I still keep in contact. In Germany, I tried to establish a few contacts, some of them didn't work out." (alum 3 2022). Maintaining an active alumni network can be challenging for several reasons. While some alumni remain proactive in staying connected with the project, staff, and scholars, others may lose touch and fail to update their contact information. In such cases, it is important to go beyond official channels like mailing lists and foster a snowball effect, encouraging alumni to share information with each other. Additionally, the network should not function as a passive mailing list for sending out calls, conference announcements, and vacancy notices. Instead, alumni should be actively involved in academic activities, such as being invited as speakers or lecturers, co-authoring articles, or collaborating on funding applications. An example of this approach is the DAAD-funded project "Advanced Training in Co-Creating Avenues for Culture and Sustainable Development" (2022–2023), where SDG Graduate School alumni, alongside staff from the University of Hildesheim and Zeppelin University Friedrichshafen, co-developed workshops that addressed participants' academic interests and methodological needs, strengthening the network for future research collaborations.

While this initiative is a step towards consolidating the SDG GS alumni network, a strategic and sustainable approach is still needed. The University of Hildesheim is only just beginning to establish an international alumni network, and with the SDG GS set to conclude in 2025, questions remain about who will be responsible for maintaining the alumni network and whether they possess the necessary intercultural competencies to do so.

This observation may lead to another aspect that is important when considering the question of how to build sustainable international academic cooperation: training students and staff in intercultural awareness.

Working Around the "Awkward Moments": Training in Intercultural Awareness

The trilateral nature of the GS continuously exposes the scholarship holders to different cultural contexts that may be unfamiliar to them. This exposure may happen through both in-person and virtual events in which the three partner institutions participate, for example the monthly online colloquia, the biannual doctoral workshops, the three-month research stay in Hildesheim, and international conferences. This not only creates an opportunity for exposure to scholars from other countries, but also an exchange of different cultural and situated knowledge, as one of the Ghana-based alumni and current project coordinator states:

> Maybe if I was just studying with a scholarship from University of Cape Coast only, I wouldn't have had certain things to do with Germans, with Nigerians. But all these things exposed me. And for me, the other side of it has to do with the introduction to other cultures and spaces. (alum 7 2022)

While this international exchange within the GS seems to represent a mere advantage for those who participate in it, the reality of a trilateral academic cooperation proves to be much more challenging. In fact, being confronted with different perspectives and cultural practices might even be perceived as a "cultural shock", as one alumnus states:

> Of course, there's a cultural shock. (…) Of course, bringing people from different angles of the world, from different cultural spheres, different mentalities, there's bound to be a cultural shock. Communication codes, moral codes are not the same, misinterpretations, so there are always those awkward moments where we don't understand. (alum 5 2022)

However, discovering that practices and perceptions might vary from one person to another, from one cultural and/or academic context to another, can open the gateways for an understanding of one another, which ultimately enriches the perspectives on academic questions and discourses. Another interviewed alumnus describes this phenomenon as a kind of mutual and two-sided learning process:

> When you look at your Ghanaian friends, you already know how their life is, you know them, you have the picture. But now you see and you are more likely to accept diversity. On another level, you also get to appreciate the fact that we all try to learn, and they have a lot to learn from each other. For instance, when we have a concept, and I talk with others and learn." (alum 3 2022)

In order to enhance this two-sided learning process, the various platforms of the project can be used at different points in time. For instance, through the participation of staff members and resource persons from different backgrounds, the doctoral workshops themselves confront the scholarship holders with different standpoints—academic, cultural, political, religious and so on. Some workshop sessions might also be given a specific focus in order to enhance a mutual understanding, for example by preparing the PhD students for their three-month research stay in Germany or to discuss different cultural approaches with regard to gender roles. At a later point of their PhD process, namely during their stay at the University of Hildesheim, the scholarship holders can benefit from further training units with regard to intercultural understanding, especially in everyday interactions, for example through introductions to German supermarkets, language courses and courses on everyday and/or structural racism in Germany. However, these courses cannot fully prevent the aforementioned "cultural shocks" from happening. For instance, some PhD candidates who visited an ethnological exhibition were quite shocked by its racist colonial depictions. These are but a few impressions shared during regular feedback sessions with board members and students highlighting the intricacies of intercultural academic exchanges.

However, with regard to former and present power hierarchies within such cooperations, which mainly stem from former colonial structures and current global inequalities, it is important to highlight that exchange within the trilateral space needs to be three-sided and well balanced, and that one side—usually from the Global North—should refrain from imposing its knowledge and project goals on the other partners. Through one of the alumni's previous citations, it becomes clear that such spaces of intercultural exchange require all the parties involved to treat each other with respect:

> Let the relationship continue, let the academic exchange continue, let me respect your culture and let you respect my culture. Let me respect your language and respect my language. Let me respect your religion… that is what I mean by sustainability. (alum 1 2022)

Beyond the mere question of mutual respect for different discourses, practices, value and norm systems, the topic of transcultural awareness in academic cooperation of course remains a question of continuous negotiation processes. Having three different 'people', as one of the alumni and current coordinators puts it, coming together, bringing their different standpoints and practices to the table, requires greater flexibility when it comes to individual or institutional ways of doing things, and establishing a system of power sharing and, ultimately, compromise. Finding middle ground between the different project partners, institutions and staff members, students and resource persons, is especially important when it comes to the challenges of conducting such an international project.

5 The Operative and Structural Challenges of a Trilateral Graduate School

This section addresses the various operative and structural challenges faced by the SDG Graduate School. Specifically, we focus on gender-based issues, decolonizing research cooperation, and managing academic life, particularly difficulties encountered in entering the writing process and managing expectations from different institutions. Here, we will first discuss the perspectives of students and alumni before adding our own experiences as coordinators.

Gender-Based Challenges: Equality, Motherhood, and Family Issues

As a GS that centers its vision around the SDGs, the partners did not just try to see this in the different projects of the scholarship holders but also sought to mainstream it within its daily practices. Hence, the project tried to ensure equal representation for the scholarships among qualified candidates. The GS also put in place checks and balances that afforded both genders the environment they needed to thrive, such as an opportunity for nursing mothers to hire nannies during workshops if required. Another opportunity put in place was the availability of medical breaks with the right support so that scholars could get the medical attention they needed. This was particularly useful in the Nigerian context when a female scholar experienced some mental health issues and needed to take a break from her studies. It is undeniable that being part of such a trilateral partnership as the GS comes with advantages and disadvantages that vary from one country to the next as well as amongst the candidates themselves. One of the Nigerian interviewees stated that:

> It was difficult for me to leave my daughter at home… move to an entirely different country to study. Sometimes I call and she is crying over the phone requesting for my presence, while other times she refuses to talk to me, thinking that I abandoned her… it was a tough time and a tougher decision. (alum 2 2022)

On the same note, another alumna recalls:

> I remember it was a huge conflict for me as I can confidently say it affected the relationship between myself and husband negatively…he didn't see why I needed a PhD and refused to support the process…a huge challenge, coming from a patriarchal society, where I needed his permission to further my studies. (alum 6 2022)

Such responses clearly show how difficulties in terms of family issues affected some of the GS scholars. While the measures described above were put into place for individual scholarship holders in need of additional support, be it financially or with regard to their role as caregivers and its compatibility (or lack thereof) with their research activities, structural measures with regard to parenthood[1] and family hardship are entirely lacking from the SDG GS (financial) structure. Thus, for future cooperation projects it would be advisable to include additional financial support for parents-to-be, and especially for scholarship holders who give birth, so that they can reconcile their care activities with their research projects, and not see their academic work jeopardized by their family obligations, or vice versa. While support could be provided through an additional child allowance within the scholarship itself, it could also be taken into consideration by offering a prolonged scholarship period to those who have given birth during the period of their scholarship.

Equal Partnerships and Decolonizing Research Cooperations

Decolonizing research cooperations has been of particular importance within such partnerships as the trilateral SDG GS. Looking at the United Nations mandate to

---

[1] We have chosen to include the term 'motherhood' in the title of this section because, within our project's context and cultural setting, referring to biological mothers as being most affected by gender-based family obligations seemed most appropriate in terms of personal experiences. However, we are aware that parenthood in general and the obligations that come with it are often reduced to the identity of biological motherhood, and that this creates an overpowering social and mental load for biological mothers, while at the same time excluding other, more invisible forms of parenthood. In general, it is advisable to have additional support measures included in international research cooperations which apply to parents and caregivers in general, regardless of their gender or genetic relation to the child.

“strengthen the means of implementation and revitalize the global partnership for sustainable development” (UN 2015), the GS has followed this mandate to the letter through methods of equal decision-making and cotutelle processes where the partner universities respect each other’s peculiarities in terms of supervision and certification. As one of the interviewed alumni states:

> …only in this project have I seen universities make real commitment and adjustments to processes of certification while upholding the home university mandate on ethics and due process… and still have things move according to plan… successfully. (alum 9 2022)

Further examples of equal partnership are seen in the biannual workshops, whose processes involve inviting resource persons such as members of academic institutions in ethnomusicology, cultural studies and other related disciplines, but also practitioners from civil society organizations in the field of peace and conflict, policy makers, stakeholders in cultural institutions, etc. from the partner home country and the full involvement of the partner university in the implementation of the workshops. It is important to state that the GS, in its efforts to achieve an inclusive learning environment, implements a trilateral mobility system. The German partners visit the Ghanaian and Nigerian partners during the biannual doctoral workshops and gain insights into the local academic and cultural contexts and, in turn, scholars from Ghana and Nigeria participate in a three-month research study exchange where they have the opportunity to immerse themselves in the educational and cultural fabric of German society. There is also an opportunity for a resource person from both West African partner universities to take up a month-long research visit in Germany and an opportunity for a one-month South-South exchange between the partner institutions in Ghana and Nigeria, where the resource person participates in the activities of the host university. In order to sustain research cooperations in the long term, the SDG GS also financially supports infrastructural measures at the partner institutions, especially with regard to the Centre for Study and Promotion of Cultural Sustainability (CSPCS) at the University of Maiduguri. This research center has been funded within the framework of the SDG GS and has been the beneficiary of important infrastructural measures, such as providing the equipment for a conference system and solar panels in order to ensure a reliable electricity supply, including during power outages.

To further strengthen the importance of partnerships between the EU and Africa, the UN held a plenary assembly session in 2020 on equal partnership, which clearly emphasized the need for high-level cooperation (Dimitriadis et al. 2020). This was deemed necessary because Africa is included on the index of one of the 20 fastest growing economies in the world. The EU document (The European Union and Africa 2021) further stated that by 2025 Africa will have the largest potential workforce globally, perhaps turning the spotlight on such partnerships that involve mutual respect and diverse cultural exchange? This presents an opportunity to foster regional value chains, sustainable local economic development and the production of homebred capacities by creating sustainable context-based knowledge production with a degree of international flair, while also potentially putting into perspective the aforementioned, mostly Western-centered definition of development and opening up the possibility of redefining it from a Sub-Saharan African standpoint.

Entering the Writing Process

One of the difficulties experienced by several candidates was and remains the writing process, or rather getting started on writing their final thesis. Some alumni were very expressive about their difficulties with the academic writing process. For instance, one alumnus told us that:

> …..I felt lost in the beginning, I was not even sure what I should write, which I can was caused by information overload, in fact I downloaded everything that had anything to do with my variables and half was useless at the end of day ….but how we go do now, man must progress (but what do I do? I have to progress) …. (alum 10 2022)

Another alumna had a similar response where she stated that "getting into the zone was a challenge, but when I got started it flowed…the writing boot camp helped us a lot, it put me in some sort of zone, people were nowhere, we had targets for each day… I took this method and applied it even after the boot camp". (alum 8 2022) The struggle of students during the writing processes is quite evident, especially within such a time-conscious project as the GS. Providing workshops and boot camps where students can learn effective writing methods (such as setting daily goals) and develop time management skills has proven a major contribution to the success of the program. There are, however, multiple factors such as obligations towards family and work, especially when students begin lecturing at their home universities, which impact their ability to concentrate on writing. Additionally, the official writing process begins at different times depending on the student's home university. Typically, students start their research before submitting the proposal that secures their scholarship. Many alumni reported conducting primary research throughout their coursework, followed by a phase of data collection, analysis, and finally, writing their thesis. The three-month research and writing stay in Germany, which occurs towards the end of the scholarship, usually coincides with the completion of data collection and the beginning of the thesis writing process. During this stay, German coordinators typically organize a program that includes writing workshops, educational excursions to German and European research institutions, and participation in major conferences. While students generally gave positive feedback about the program, we significantly reduced scheduled events to prioritize supervision and self-organized writing time. Balancing this program in response to the students' challenges remains a key focus for planning the final cohort's stay in 2025.

Processing Workloads from Different Institutions and Managing Expectations

PhD students at the GS face several challenges, particularly in managing the workloads from two different institutions and meeting the diverse expectations set by the school. These challenges arise at multiple levels, including the administrative requirements imposed by partner universities and the DAAD, the funding body, as well as the varying individual expectations from supervisors, which can differ significantly in terms of their respective research areas, academic traditions, and overall standards. For instance, students pursuing a PhD under a cotutelle arrangement must satisfy the requirements of both universities involved. At their African home university, students are not required to publish their thesis to meet PhD requirements.

Instead, they must participate in multiple colloquia, defend their thesis at various stages of completion, and respond to new committee requirements before submitting their thesis. In contrast, the requirements in Germany do not include multiple defenses before submission, but students must publish their thesis within two years of the final defense.

Despite these challenges, the cotutelle process has gained popularity among PhD students at the GS, particularly those in the second and third cohorts. The feedback from our students on this issue varies greatly, although their overall assessment is positive:

> The major challenge was having to meet the requirements of the different universities which were a bit and in some cases to a very large extent different. However, the experience was rewarding in the sense that it made me work extra hard to meet the requirements of each university, making me a more resilient scholar. The binational degree is also gaining more popularity, and I count myself privileged to have it. (alum 4 2022)

The challenges faced by our students are thus multifaceted and operate on various levels. While some challenges are individual, the majority are structural, particularly those related to gender-specific issues and historical power imbalances between the Global North and South. This requires us, as coordinators, to reflect on and potentially adapt our approaches to teaching, organizing, and managing the GS, taking the feedback of students and alumni seriously. This already leads us to the challenges we faced as local coordinators in Germany, Ghana and Nigeria, which can be summarized by briefly discussing two larger topics: Managing and communicating different requirements for the students, and securing additional funding sources to assure the program's sustainability.

Coordinator's Perspectives

One of the key challenges for us as coordinators of the GS is communicating and managing different expectations from the different entities involved in this cooperation. First and foremost, the third-party funding from the German Academic Exchange Service (DAAD) through the Federal Ministry for Economic Cooperation and Development (BMZ) requires setting an agenda. This consists of project goals and milestones that have to be put into practice and closely monitored. While the main project outcome is, of course, the financial and intellectual support of both MA and PhD scholarship holders at the University of Cape Coast and the University of Maiduguri, the program actually contains a wide range of expected project outcomes that we have to attend to as coordinators, such as sustaining the infrastructure of the program, the mobility of scholars, and the development of a long-term RDM strategy. Each item on the program's agenda is connected to a certain budget which is set by the project's financial plan.

At the same time, the needs of our students and staff at the three partner universities may vary over the time of the project, and project goals may shift or seem unrealistic to fulfill for various reasons or within the given budget, be it for administrative reasons, economic crisis or inflation, or logistical and operative changes. Our role as coordinators is to mediate between the determined project outcomes that form the basis of the DAAD third-party funding and which are oriented towards the sustainability of the trilateral cooperations, and the actual and ever-transforming needs, expectations and conditions at each of the three partner universities. This form of mediation requires close communication between the different entities of the scholarship holders and staff in Ghana, Nigeria, and Germany alike, and the responsible resource persons at the DAAD. It is only through a two-way communication that transformative processes within the project can truly take place and be accompanied accordingly.

Another significant challenge lies in securing additional funding sources beyond the support provided by the DAAD. While we have identified several potential funding bodies that could contribute to specific aspects of the GS, such as workshops and collaborative events and projects, finding sustainable funding to support the entire structure of the GS remains a major hurdle. While this funding challenge may be common among many GS currently, our GS faces specific obstacles. Firstly, our emphasis on culture does not yield direct economic or technological outcomes, which limits the marketability of our outcomes. Secondly, the cultural sector, particularly "the arts", suffers from chronic underfunding. This issue became particularly prominent during the pandemic, when cultural workers faced immense struggles, and the arts were deemed non-essential by governments and administrations in Germany, Nigeria, and Ghana alike.

Lastly, finding suitable supervisors for the cotutelle at the German university posed a challenge for some of our students. Especially when it comes to institutional differences in understandings of the defining concepts of the GS, i.e. culture, development, performance and sustainability, as described in the beginning, finding suitable supervisors for the cotutelle process at the German host university can be challenging. As stated earlier, "culture" often translates to "the arts" in German academia, which in some cases required us to find supervisors in other faculties of the university. This again means that students (and GS staff) have to consider specific requirements for the cotutelle process.

Building on the GS's commitment to decolonizing research practices and fostering gender equality, we have shown how these principles are actively embedded within the GS's educational framework. This section has highlighted the challenges and successes in promoting a collaborative academic environment that bridges diverse cultural and academic expectations across the partner universities in Germany, Ghana, and Nigeria. We now shift the discussion to the promotion of sustainability and the integration of the SDGs within the GS. Here, the GS's strategic efforts to embed sustainability across its organizational structures, research practices, and partnerships are examined in depth, offering key insights into how the program fosters a critical and locally informed understanding of sustainability.

6 The Promotion of Sustainability and SDG Within the GS

In this last section, we explore how the GS integrates sustainability and the SDGs into its operations and partner institutions. The GS actively promotes sustainability and the Sustainable Development Goals across various levels. Primarily, it strives to establish sustainable frameworks within its own organization and the host universities involved.

In its second phase of funding, the GS has also implemented a dedicated data management strategy. Each partner university employs a research data management expert to ensure the long-term viability of data generated by GS members. The research data management team develops strategies and guidelines to achieve the following objectives: (1) ethical handling of sensitive data, (2) broad accessibility and availability beyond individual researchers, and (3) countering extractivist logics by housing GS-generated data sets at each partner university.

Beyond this crucial conception of a shared, sustainable research data management, the GS promotes and implements sustainable concepts with a focus on institutional integration at partner universities and the search for external funding opportunities. We conclude this section by providing 4 key insights into promoting sustainability between the Global North and South.

Institutional Integration

The GS has developed its own organizational structures within the host university. At the University of Hildesheim, the GS operates within the Center for World Music and is jointly managed by the center's staff and the staff of the UNESCO Chair On Cultural Policies for the Arts and Transformation at the Institute for Cultural Policy. Hosting and administering the GS contributes to the development and enhanced integration of the Center for World Music within the university's structure. It also helps to institutionalize positions for administrative staff and researchers, while fostering greater awareness of fundamental cultural sustainability issues within the university and beyond.

At the University of Cape Coast, the Global Sustainability (GS) initiative and its members have been successfully integrated into existing structures, particularly within the Institutes of Development Studies, Geography and Regional Planning, and the Department of Music and Dance.

In Maiduguri, the funding provided to the GS has facilitated the establishment of the Centre for the Study and Promotion of Cultural Sustainability (CSPCS). This center serves as the host for the GS and creates a vibrant intellectual environment for scholars engaged in cultural studies at both national and international levels. The CSPCS conducts research, documentation, preservation, interpretation, and digitization of archives and collections. These activities aim to enhance the academic and social services offered by the center and to increase the accessibility and viability of cultural heritage for future generations.

Sustainability Beyond Funding

All partner universities are actively working towards the establishment of fully self-sufficient and sustainable structures beyond the funding phase provided by the German Academic Exchange Service (DAAD). A crucial aspect of this endeavor is the development of strong international scholarly networks and partnerships. Key to achieving this are the aforementioned biannual workshops hosted alternately at the University of Cape Coast and the University of Maiduguri, which bring together national and international experts from various disciplines, including ethnomusicology, cultural studies, conflict studies, and cultural policy.

National and international conferences, particularly the biannual conferences held in Maiduguri, have garnered significant media coverage, thereby raising awareness among the interested public regarding sustainability and the role of culture in mediating, creating, and maintaining sustainability. Moreover, students, alumni, and scholars associated with the GS frequently contribute to academic papers and conference presentations that address issues pertaining to culture and sustainability. These contributions greatly add to the visibility and recognition of sustainability as a crucial topic within global cultural studies.

Focus on Specific SDGs

The GS's primary focus lies on SDG 16 (Peace, Justice and Strong Institutions). The reason for this specific focus can be attributed to the devastating impact of the Boko Haram insurgency, particularly in Nigeria's northern and eastern regions, and notably Borno State where the University of Maiduguri is situated (Riva 2020). Additionally, while SDG 16 remains central, our students have often engaged with issues that intersect with other SDGs, notably SDGs 4 and 5, which prioritize Quality Education and Gender Equality respectively. Examples of such intersections can be found in the footnotes highlighting our alumni's work.

However, it is important to note that the GS "Performing Sustainability" initiative does not merely promote sustainability but also aims to cultivate a critical and comprehensive understanding of the SDGs within a broader, globalized, and postcolonial framework. The majority of our students reported having had limited prior comprehension of the concept of sustainability or the SDGs. It is through the GS's curriculum and workshops that they have developed a deeper understanding, particularly in relation to culture. As mentioned in an earlier section of this paper, sustainability lacks a singular conception, and there is no prevailing understanding of the role of culture in sustainability. Instead, the GS encourages students to critically analyze and explore various theoretical perspectives on (cultural) sustainability within the context of their own research. This critical engagement has led to the emergence of multiple aspects that contribute to a more enriched and nuanced understanding of sustainability, moving beyond the notion of cultural conservation, and encompassing the three (or four) pillars, as addressed by Wu, Fan and Chen (ead. 2015), among others. Through interviews with coordinators, students, and alumni, we have identified four interrelated insights into sustainability and sustainable processes that have become apparent throughout the duration of the GS program.

Insights Gained: Fieldwork and Local Contexts

The first and arguably the most significant insight emerged early on during fieldwork experiences. As part of the GS program, PhD students in Ghana and Nigeria are required to undertake a year of coursework at their respective host universities prior to conducting field research. During this initial year, students delve into relevant theories and concepts pertaining to their specific fields of study. The GS curriculum in this phase focuses on research methodologies, as well as core concepts and theories related to culture, development, sustainability, and conflict and peace. Following this preparatory period, students embark on fieldwork, which often involves direct engagement with diverse cultures and ethnic groups. Within these ethnographic encounters with the "Other" (Said 1978), students are tasked with situating themselves within new social and cultural contexts and navigating the diverse and sometimes conflicting agendas of the individuals they collaborate with. In these encounters, sustainability implies an approach that starts from the grassroots, taking into serious consideration the needs and perspectives of the people involved. Hence, a heightened sensitivity to power dynamics, neocolonial structures, and individual positionalities becomes imperative.

Insights Gained: Local Understandings of SDGs

The second insight closely relates to the first and emphasizes the necessity of creating, recognizing, and mediating local understandings of sustainability and the SDGs. One of the students aptly expresses this notion by stating:

> The way cultural sustainability is viewed from the Global North is different from the way the Global South views it. For instance, how we as people may have certain social norms that forbid us from doing certain things, for instance "if you have this, you have to go to the farm and if you have this, you should not go to the farm". So, these are the ways in which we preserve the environment. You have limited time to go to the woods and fell down trees, and once you break those rules you have a penalty to pay. If you don't want to pay you have to stick to those rules. These are the ways local communities were preserving the environment which are unique to them. Now, if you want to look at it, they [people from the Global North] tell you that they are backwards, uncivilized and their culture is barbaric. And they tell people who do not to go to the farm on this day and ask why, you are supposed to work, to look at it from an economic perspective, and they don't know why they are doing that. They are doing that so that they can preserve these things, but looking at it from an economic, Northern perspective it is wasting time. So understanding the way people do things in Africa is very important for our understanding of sustainability. (alum 2 2022)

Social and cultural institutions, which may not be immediately recognized as contributors to sustainability, often form an integral part of the broader cultural framework that supports local practices of sustainable living. In some instances, researchers are required to identify and understand the local manifestations of the SDGs, thereby assuming the role of cultural mediators. As shared by an alumnus:

> It is no longer just about not depleting resources for the future generations, but it's also about not prescribing development to people, understanding their development as the final cultural alliance. What is development for some may not be for others, especially when the usefulness is in parenthesis, because the cultural values of the target beneficiary were not taken into consideration when the projects or initiatives were conceptualized." (alum 5 2022)

Insights Gained: Sustainability as a Process

The third insight derived from the preceding two highlights the importance of perceiving sustainability, particularly in the form of the SDGs, not as a final destination but as a process-oriented endeavor. This perspective aligns with our previous concerns regarding the funding structure of the GS and the predominant focus on "development", which can undermine efforts to foster equitable bilateral and trilateral cooperation between the Global North and South. Our students and alumni have emphasized that the creation and exchange of knowledge, both among scholars and between researchers and research participants, should be viewed as an ongoing and often open-ended process. This perspective places greater emphasis on understanding the methodologies and rationales involved in working towards the SDGs and on the ways in which scholars navigate diverse agendas and varying local interpretations.

> The trilateral nature of the project made me look at concepts and theories from three perspectives: German, Ghanaian and Nigerian. I had the opportunity of learning from international scholars which gave me a broader understanding of concepts and theories from diverse angles. Such a setting made me appreciate diversity a lot which is an integral part of the SDGs. As an African scholar, I became more critical about what I accept as authentic. I became especially critical about questioning how concepts and theories are explained to us and how we accept them. (alum 4 2022)

Insights Gained: Community Engagement

The fourth insight pertains to the importance of community engagement. This principle holds true across all aspects of research, teaching, and the dissemination of knowledge. Integrating community engagement into the principles of the GS generally means to aspire to create academic knowledge and research practice in exchange with civil society. More specifically, community engagement takes place in the GS through the internships of the scholarship holders in civil society organizations, carrying out field research for the MA and PhD theses in local communities, and engaging with said communities and civil society organizations during the biannual doctoral workshops.

These insights naturally lead us to the question that preoccupies the GS and individual scholars: How can we conduct sustainable research? The GS has actively pursued and continues to explore various avenues in this regard. An essential step is to involve the communities with whom we conduct research in every phase of the research process. This inclusive approach recognizes the importance of collaboration, shared decision-making, and mutual learning between researchers and the communities they engage with.

Another vital aspect is ensuring the accessibility and re-usability of research data and findings, thereby challenging the extractive practices that have historically stemmed from colonial times. The GS endeavors to move away from a one-sided acquisition of knowledge and towards a more equitable and collaborative approach that fosters knowledge exchange and reciprocity. Making research data and findings openly available can empower communities and contribute to the broader goals of sustainability.

7 Conclusion

In summary, the GS has made significant strides in promoting sustainability and the Sustainable Development Goals (SDGs) on multiple levels. It works towards establishing sustainable structures within itself and its host universities. The GS has implemented its own data management strategy, ensuring the ethical treatment and accessibility of data generated by its members. Through biannual workshops, the GS has facilitated knowledge exchange and networking among students, experts, and GS members from diverse fields, enhancing their academic profiles and fostering international collaborations. The GS's curriculum and workshops have played a crucial role in cultivating a critical understanding of sustainability within a globalized and postcolonial framework, challenging students to critically analyze and probe different theories of cultural sustainability. Through community engagement and the sharing of research data and findings, the GS aims to conduct sustainable research that fosters inclusivity, reciprocity, and countering extractivist logics derived from colonial times. These efforts reflect the GS's commitment to sustainable practices and its broader goal of advancing cultural sustainability and the SDGs on a global scale.

We have shown that the challenges faced by our students are multifaceted, encompassing both individual and structural levels. Predominantly, these challenges are structural, particularly those associated with gender-specific issues and historical power imbalances between the Global North and South. As coordinators, this necessitates a reflection on and potential adaptation of our approaches to teaching, organizing, and managing the GS, taking student and alumni feedback seriously. The challenges we faced as local coordinators in Germany, Ghana, and Nigeria can be summarized by discussing two major topics. First, managing and communicating different requirements for students, such as balancing milestones set by the DAAD with those from home universities. Second, securing additional funding sources beyond the DAAD's support remains a significant hurdle. Although we have identified potential funding bodies for specific aspects of the GS, finding sustainable funding for the entire structure is challenging. In addition, the GS has inspired the partner institutions to envision ambitious goals for future collaborations, particularly in establishing long-term, equitable partnerships. A key aspect of this decolonizing effort would be the shared administration of available funding. Additionally, it is essential that PhD exchanges between partner universities become normalized within these institutions, with administrative support and consideration of differing academic realities, such as variations in salaries and teaching obligations. Furthermore, it remains crucial to contribute to the development of sustainable academic infrastructures, like the CSPCS at the University of Maiduguri, ensuring their eventual independence from Western funding.

In conclusion, we can respond to the various challenges that we continuously face within the trilateral cooperation of the GS by applying different strategies, as discussed in the first part of the main chapter of this article: by exchanging and negotiating differences, bringing together different institutional settings, putting our focus on capacity building within the network, creating international networks, and training scholarship holders and staff in intercultural awareness. All these strategies help to create an academic platform for different voices, however hard it may be to implement them completely in a project that is not only a cooperation between three different institutions but also involves third-party funding. The GS can thus be considered a space within which we do our best to make room for respectful, open and decolonized exchanges and debates, while also using it as a platform to create networks within and outside the GS and allowing for a sense of intercultural awareness. While the current funding of the program is limited and will end on December 31st, 2025, applying strategies for capacity building within the network constitutes an attempt to sustain both the GS as an academic space and the academic and social contacts within it.

References

Austin JL (1962) How to do things with words. Oxford University Press, Oxford

Bauman Z, May T (1990) Thinking sociologically. Blackwell Publishers Ltd., Oxford

Brundtland GH (1987) Report of the World Commission on Environment and Development: our common future. United Nations General Assembly document A/42/427

Cenere S (2022) Making in the making; performing new forms and spatialities of production. FrancoAngeli Publications, Milan

Clark EH (2021) Introduction: audibilities of colonialism and extractivism. World Music 10(2):5–20. Retrieved from https://www.jstor.org/stable/27095336

Dimitriadis D, Slavona D, Wagnsonner T (2020) Opinion of the European Economic and Social Committee on 'EU and Africa: making an equal development partnership a reality based on sustainability and common values'. Off J Eur Union 429:105–113. Retrieved from https://eur-lex.europa.eu/legal-content/EN/TXT/PDF/?uri=CELEX:52020IE1715&from=EN. Accessed 27 Sept 2024

The European Union and Africa (2021) Towards a partnership 'between equals'? Retrieved from https://www.plataformaongd.pt/uploads/subcanais2/the_european_union_and_africa_towards_a_partnership_between-equals_englishversion.pdf. Accessed 26 Sept 2024

Heinicke J (2023) Von der kolonialen Entwicklung zur postkolonialen Transformation: Kulturpolitik als Weltaufgabe. In: Crückeberg J et al (eds) Handbuch Kulturpolitik. Springer, Wiesbaden

Lettau M, Mtaku CY, Otchere ED (eds) (2023) Performing sustainability in West Africa: cultural practices and policies for sustainable development. Taylor & Francis, Abingdon

Ododo SE, Tijani AI, Vogels R (eds) (2020) Cultural sustainability, performance and the sustainable development goals in time of crisis. Centre for the Study and Promotion of Cultural Sustainability, University of Maiduguri, Nigeria and Centre for World Music, University of Hildesheim, Maiduguri/Hildesheim

Phelan P (1993) Unmarked: the politics of performance. Routledge, London

Purvis B, Mao Y, Robinson D (2019) Three pillars of sustainability: in search of conceptual origins. In: Sustainability science, vol 14. Springer, New York, pp 681–695. https://doi.org/10.1007/s11625-018-0627-5

Riva N (2020) Academia under attack—Wissenschaft unter Beschuss: accounts of the Boko Haram insurgency at the University of Maiduguri. Berichte über den Boko Haram Aufstand an der Universität zu Maiduguri, Georg Olms Verlag, Hildesheim

Said EW (1978) Orientalism. Pantheon Books, New York

Schech S, Haggis J (2000) Culture and development: a critical introduction. Blackwell Publishers Ltd., Oxford

Sen A (1999) Development as freedom. Alfred A. Knopf, New York
Taylor D (2016) Performance. Duke University Press, Durham/London
Tylor EB (1871) Primitive culture: researches into the development of mythology, philosophy, religion, art, and custom. John Murray, London
Williams R (1958) Culture and society 1780–1950. Chatto & Windus, London/Harper & Row, New York
Wu SR, Fan P, Chen J (2015) Incorporating culture into sustainable development: a cultural sustainability index framework for green buildings. Sustain Dev 24(1):64–76. https://doi.org/10.1002/sd.1608

Websites

United Nations (2024) UN SDG Goal 17. https://sdgs.un.org/goals/goal17. Accessed 26 Sept 2024

Lea Frauenknecht is an academic associate at the Institute for Cultural Policy at the University of Hildesheim, Germany. Since November 2022, she is a coordinator of the DAAD-funded SDG Graduate School "Performing Sustainability. Cultures and Development in West Africa". Since April 2025, she is also a research fellow in the EU-funded project INTRACOMP (Intercultural and Transcultural Competence Through Collaborative Cultural Expression). Her research interests include critical cultural mediation and critical cultural theory, post- and decolonial discourses on sustainability, gender and queer theory as well as social movements.

Martin Ringsmut is a postdoctoral university assistant in ethnomusicology at the University of Vienna and former coordinator of the SDG Graduate School *Performing Sustainability*. His research focuses on the music of Cabo Verde and the Sinti and Roma in Europe, exploring aspects of cultural identity, memory processes, postcolonialism, and social spatiality.

Naomi Andrew Haruna is a Senior Lecturer in Industrial Design Department at the University of Maiduguri, specializing in Cultural Sustainability and Visual Communication. She coordinates the Nigerian KOSTIMA project which focuses one hegemonic voices in achieved BMDP and the SDG Graduate School Project. Her research focuses on culturally based visual representations for peace promotion among internally displaced persons in Northeast Nigeria, with interests in Peace Media, Development Communication, and Gender Studies. She is a fellow with the Salzburg Global Institute and a member of the Society of Nigerian Artists.

Sabina Appiah-Boateng is a Research Fellow at the Department of Peace Studies, University of Cape Coast. She holds a Ph.D. in Development Studies from the University of Hildesheim, Germany, and the University of Cape Coast, Ghana. She is also a fellow of the African Humanities Program of the American Council of Learned Society and the Next Generation Social Sciences Research Council in Africa. Her research interests include peace and conflict studies, mental health, cultural sustainability and gender and development. She has consulted for both local and international organizations, including USAID, UNDP, WANEP, European Union, National Peace Council, and AngloGold.

Responding to Contextual Challenges in Urban Africa: A Collaborative Postgraduate Approach for Sustainable Development Through Urban Management—The Wits-TUB-UNILAG Urban Lab

Mfaniseni Fana Sihlongonyane, Lucas-Andrés Elsner, Avhatakali Sithagu, Marie Huchzermeyer, and Taibat Lawanson

Abstract The African continent has been witnessing one of the highest rates of urbanisation globally, albeit uneven. While rapid urbanisation creates opportunities, it poses seemingly insurmountable challenges exacerbated by global threats such as climate change. This chapter discusses how an international North-South and South-South collaboration sought to respond to the African urban conundrum. This German-funded collaboration as of 2016 has been between the University of the Witwatersrand in Johannesburg (Wits), South Africa and the Technical University Berlin (TUB), Germany, with the University of Lagos (UNILAG), Nigeria, joining as of 2021. The chapter discusses both phases of what became the Wits-TUB-UNILAG Urban Lab and the introduction of masters degrees in urban management at Wits and UNILAG. From a review of urban change on the African continent, the chapter discusses critiques of how the urban Sustainable Development Goal and

Avhatakali Sithagu passed away on 13 October 2024. She contributed to a draft of this chapter.

M. F. Sihlongonyane · A. Sithagu · M. Huchzermeyer
School of Architecture and Planning, University of the Witwatersrand, Johannesburg, South Africa
e-mail: Mfaniseni.sihlongonyane@wits.ac.za

M. Huchzermeyer
e-mail: Marie.huchzermeyer@wits.ac.za

Present Address:
L.-A. Elsner (✉)
Habitat Unit, Technische Universität Berlin, Berlin, Germany
e-mail: lucas.elsner@posteo.de

T. Lawanson
Centre for Housing and Sustainable Development, University of Lagos, Lagos, Nigeria
e-mail: tlawanson@unilag.edu.ng

L.-A. Elsner
Sustainable Cities and Climate Change, HafenCity Universität Hamburg, Hamburg, Germany

C. Brandt et al. (eds.), *Implementing Sustainability and Internationalization in Higher Education*, World Sustainability Series, https://doi.org/10.1007/978-3-031-97837-1_5

the New Urban Agenda respond. It sets out how inequalities in knowledge production and capacity in the context of the legacy of colonial planning, with the field of urban management having emerged in this context. Against this backdrop, the paper presents the long-term gains of the approaches adopted, including core themes for the curriculum.

1 Background: Taking Stock of Urban Change in Africa

Johannesburg and Berlin are starkly contrasting cities. The former is deeply segregated, sprawling and motorcar oriented, a spatial model that traps impoverished households in neighbourhoods from which economic opportunities are difficult to reach. The latter is compact and threaded with efficient public transport. Yet these two cities share a troubled past, historic moments of reconciliation, yet persistent legacies. The ongoing search for solutions to urban challenges across Johannesburg and Berlin has long inspired student and staff exchanges. As a result, the Habitat Unit in the Architecture School at Technische Universität Berlin (TUB) and the School of Architecture and Planning (SOAP) at the University of the Witwatersrand (Wits) in Johannesburg have collaborated on an incremental, ad hoc basis since the late 1990s. The call in 2016 by the German Academic Exchange Service (DAAD) with funds from the German Federal Ministry for Economic Cooperation and Development (BMZ) for SDG Graduate Schools in the global South in partnership with German universities presented the opportunity for the Wits-TUB collaboration to consolidate, though with some initial anxiety, given the scale of the SDG Graduate School initiative.

With a successful application, which epitomised the meeting of minds with common interests for collaboration towards urban transformation and sustainable development through postgraduate curriculum, the cooperation set off as the Wits-TUB-UNILAG Urban Lab in October 2016. The objective was to make a positive impact through human capacity building to tackle and mitigate the poignant urban challenges on the African continent. During the proposal writing, funding application and the project design, the team sought to establish a horizontal and equal North-South collaboration.

In an effort to break down the North-South binary, the application for the second five-year phase of the project (2021–2025) involved a third partner, the Centre for Housing and Sustainable Development (CHSD) in the University of Lagos (UNILAG) in Nigeria. CHSD was already well networked with the Habitat Unit at TUB and was in the process of establishing an urban management masters degree. Therefore, the largest mega-city on the African continent, Lagos, promised a rich site for partnering, exchange and exploration of African urban problematics and responses in what was to become the Wits-TUB-UNILAG Urban Lab. The second phase of the project also incorporated a stronger emphasis on research, support to early career researchers and urban exchange. This was facilitated, among others by a webinar series, by horizontally co-editing and co-producing the book *Everyday*

Urban Practices in Africa: Disrupting Global Norms (Appelhans et al. 2025) and hosting an 'African Urbanisms Conference at Wits in October 2024.

Invariably, urban management formed a central thread of the cooperation, embedded within student and staff exchanges and mobilities, workshops and summer schools and an extensive bursary programme. This interdisciplinary field was already taught for close on two decades at postgraduate level at TUB to address "the currently most pressing challenges faced by urban conurbations" globally (TUB n.d.). At Wits, curriculum development through the project was to lead to the introduction of the urban management field at master's level, which would respond "to challenges associated with rapid urbanisation process[es]" on the African continent (Wits n.d.). These rationales were also reflected in the establishment of the Masters in Urban Management at UNILAG.

This chapter reflects on the approaches adopted by three institutions to foster sustainable urban development in African cities. The chapter starts with background on the nature of urban change and resulting challenges on the African continent. It then briefly discusses how the contemporary global agenda including the Sustainable Development Goals (SDG) seek to address urban challenges, and the academic critiques this has elicited. This brings the discussion to the role of universities in building critical capacities, and the limitations they face on the African continent. With a shift from modernist colonial urban planning, which continues to shape African cities, to the introduction of urban management, higher education institutions are presented with a new role. However, global, North-South inequalities in knowledge production and in access to academic publications places a significant limitation on pedagogic endeavours. This brings North-South collaboration into the discussion. The chapter then introduces urban management as an interdisciplinary field and the collaborative steps that the project team took to explore core themes for an urban management curriculum. The chapter analyses the establishment of a masters offering in urban management at Wits before briefly turning to the same at UNILAG, while also discussing the early career scholars development initiative. The conclusion considers how different needs, requirements and conditions were balanced and navigated in the project and critically reflects on the project's advantages, opportunities and constraints within the parameters of the funding available.

2 Background: Taking Stock of Urban Change in Africa

Rapid contemporary urban expansion on the African continent is an undisputed fact that has driven urban change. Given the inadequate preparation for this change due to deep-seated political and economic conditions, and the intersection with climate change, the overall discourse on African cities is one that cries out for systems and mechanisms that would secure a more equitable and secure African urban future. The growth in urbanization varies across the continent, ranging from the already heavily urbanized North Africa (47.8%) to the least urbanized sub-Saharan Africa (32.8%) (AfDB 2012). The enormous speed at which Africa's cities are growing is linked to key development trends, most prominently the accelerating economic and population growth, increasing migration from rural to urban areas, and the youth bulge (AfDB 2012; UNCTAD 2018). Land delivery is a core component of urban expansion. Rapid urban growth has often occurred where indigenous knowledge systems are used to allocate land rights and define property boundaries, especially in peri-urban areas under traditional authorities and through informal channels (Durand-Lasserve 2004; Kombe 2005; Güneralp et al. 2018; Akaateba et al. 2018). Across different African cities and countries, informal land delivery occurs through a variety channels ranging from individual, family, cooperative, customary, to neo-customary land supply (Durand-Lasserve 2004; Rakodi 2007; Ikejiofor 2009; Agheyisi 2012). In particular, "neo-customary practices … rooted directly or indirectly in the custom", as a form of the de facto land delivery system and practice, is widely attractive because these "are known to be more flexible and less bureaucratic, without counting their shorter delivery time, lower transaction costs, and adequate tenure security to motivate housing investment" (Durand-Lasserve 2004, pp. 3, 12).

The process of urbanisation has led to an increase in the number of cities across the global South. On the African continent, the total number of cities has souled from 3300 in 1990, 7600 by the beginning of the 2020s, while the "cumulative population has increased by 500 million people" (OECD/UN ECA/AfDB 2022, p. 4). The African continent hosts seven 'mega-cities', those with populations of 10-million and above (Güneralp et al. 2018), These are Cairo (Egypt), Lagos (Nigeria), Kinshasa (Democratic Republic of Congo), Accra (Ghana), Khartoum (Sudan), Nairobi (Kenya), and in South Africa, the agglomeration made up of the adjacent metropolitan municipalities Pretoria, Johannesburg and Ekurhuleni (*ibid.*). Güneralp et al. (2018) expect that cities such as Dar es Salaam (Tanzania) and Luanda (Angola), will become megacities by 2033 (*ibid.*). This rapid increase in urban populations has meant that peri-urban areas are growing much faster than formal urban centres, a process that is facilitated by the forms of land release mentioned above. This, in the absence of affordable and formal alternatives, many who migrate into Africa's cities are moving into unplanned settlements where there is lack of provision for public amenities and social services (Commins 2018). The surging demand for living space in urban areas also consumes arable land and green spaces, while inadequate infrastructure as well as regulation leads to the pollution of water sources and compromising air quality; rising crime levels are a further vice associated with rapidly growing cities (Güneralp et al. 2018; Abbey 2020).

The escalating urban population in African cities and towns has also meant people are exposed to a wide range of disasters. Economic losses and human sufferings from drought, desertification, locust infestation, outbreaks of infectious diseases, epidemics and armed conflicts are the dominant natural and human-made disasters that have rendered people in African countries vulnerable (Nur 1999). The African continent is exposed to disaster risk from various natural causes, particularly those arising from hydrometeorological hazards (UNDRR 2008). These are caused by extreme meteorological and climate events, such as floods, droughts, hurricanes, tornadoes, landslides, or mudslides (Wu et al. 2016).

Hydrometeorlogical hazards occur at short intervals (Huq et al. 2007). The iconic millennial floods in Mozambique, which affected 4.5 million people and displaced 0.5 million (Brouwer and Nhassengo 2006), were followed by flooding in Algeria in 2001 in which around 900 people lost their lives and approximately 45,000 others were adversely affected (Huq et al. 2007). A year later, heavy rain caused floods as well as mudslides, forcing large scale home evacuation across East Africa (*ibid.*). Climate change, causing heavier incidents of rainfall, heat and wind, is implicated in the frequency and severity of such incidents while also driving populations into cities with hopes for a climate resilient base, but cities themselves are poorly prepared, and their adaptation strategies largely ignore the needs of vulnerable migrants (Tietjen et al. 2023). In 2021 and 2022, South Africa experienced its most catastrophic climate change related events yet recorded and displayed its unpreparedness. A devastating wildfire unfolded in Cape Town in 2021 leading to the destruction of 600 hectares of natural vegetation, key infrastructure and valuable historic buildings and library collections at University of Cape Town; estimates of the damage are in the order of R1 billion (USD 55 million) (Liu et al. 2023). A year later, floods in Durban (City of Ethekwini in KwaZulu-Natal) were reported to have destroyed more than 12,000 homes across the socio-economic spectrum but hitting informal settlements particularly hard, as well as critical infrastructure; estimates of the damage are in the order of R40 billion (USD 0.77 billion) (Mudefi 2023).

Rapid urban growth is accompanied by what has been termed 'the urbanisation of poverty' referring to the increase in percentage number of people living in poverty in urban areas (Ravallion 2001; Vives-Miró 2022). With this comes an increase in the share of Africa's urban residents living in slum-like conditions, an aggravating source of fragility (Commins 2018). This points to structural causes such as income inequalities, with the highest rates on the globe being found in Southern Africa (Sulla et al. 2022), and inequalities in power, reflected in different economic, social, legal, and political capabilities and precipitating a different kind of 'disaster' (Homan 2003). Thus, the death of hundreds of thousands of children in Niger every year, often for treatable conditions could qualify as a disaster situation (Eriksson 2007).

The predominantly negative account of urban change in African cities has long informed debates and initiatives in the United Nations Human Settlements Programme (UN-Habitat) and its forerunner, the United Nations Centre for Human Settlements (UNCHS (Habitat)), headquartered in Nairobi since its inception in 1977. We turn, in the next section, to the contemporary policy frameworks that UN member states have agreed on and mandated UN-Habitat with supporting. In recent years, these have departed from earlier 'habitat' agendas. Cities and mega-cities have become the focus of attention, along with a shift from treating them predominantly as problems to recognising in them the potential for innovation, adaptation and mitigation. This, in turn, has sparked debate from a variety of perspectives and has implications for how solutions are conceived and ultimately how capacity-building higher education curricula are designed.

3 Responding to Urban Change in Africa: SDGs, the New Urban Agenda, Higher Education and Knowledge Production

The policy frameworks and targets that UN members states agreed to in 2015, including the 17 SDGs, the Paris Agreement on Climate Change, Addis Ababa Action Agenda on Financing for Development, and the Sendai Framework on Disaster Risk Prevention, in a direct or indirect manner, have implications for urban areas and cities. The New Urban Agenda (NUA), agreed to at UN-Habitat Habitat III summit in Quito, Ecuador a year later, translates these commitments into an urban agenda. However, its text was reached, not through a deductive or linear process, but through intense negotiation around interpretations, emphases and state obligations. Therefore, one cannot consider the contemporary urban agenda, and its implications for the African continent, without understanding these debates, some of which reach back decades before 2015 and therefore have long had the potential to inform curriculum development at higher education institutions.

The International Sustainable Urban Development Agenda and Its Critique

The new approach to understanding cities as places where solutions to sustainable development challenges can be generated is most evident in the Goal 11 of the SDGs, which the UN abbreviates as 'Sustainable Cities and Communities' (UN-Habitat 2018). African governments played a relevant role in negotiating the SDGs, and since their adoption, the SDGs have proven to be influential across Africa. Many national governments have committed to the goals and have incorporated the SDGs into national and urban development strategies. Between 2016 and 2021, 46 out of 54 African countries submitted voluntary national reports on their efforts to implement the SDGs (Croese 2022).

The NUA presents a more detailed and nuanced vision of how to achieve the SDGs in cities, highlighting, among other emphases, the importance of urban and territorial planning (Watson 2021). The NUA was endorsed by UN member states. However, by the end of 2023, only around 20% submitted reports on their implementation of the NUA by the end of 2023; nevertheless, African countries were strongly represented among those (Diaz-Sarachaga and Sarachaga 2024), indicating buy-in into the NUA on the continent.

Although both the SDGs and the NUA explicitly address issues of the global South, and related academic and policy discourses have found their way into these agendas, both agendas have been subject to criticism. The SDGs have been challenged for following a growth-oriented idea of development (albeit inclusive growth), rooted in Eurocentric perspectives, which contradict social and environmental concerns and complicates the agendas' applicability in Southern contexts (Eisenmenger et al. 2020; Nagati et al. 2023). Despite the widely held view that SDG 11 represents a major step forward, commentators also note contradictions between the different sub-goals and identify shortcomings in terms of a reductionist understanding of urban challenges and governance in Southern cities (Aust et al. 2018).

Similar concerns have been raised about the NUA, with critics pointing out that its universalist understanding of cities and urbanisation is largely based on debates and experiences from the global North, and that the proposed techno-managerial approaches are inadequate to address the dynamics of urban development in Africa (Watson 2021; Sihlongonyane 2025); Kaika 2017) outlined in Sect. 2. The NUA also promotes a state-centric idea of urban governance that does little justice to the complex realities of current urban development processes, which involve a variety of non-state actors (Watson 2016). Further, while the localisation of the SDGs in African cities became a predominant preoccupation both in practice and in academic debates, often in the context of respective funding programmes, the overarching goals are seldom questioned; the idea of development that these agendas implicate remains mostly unchallenged (Huchzermeyer et al. 2025). Academia has been crucial not only in putting forward these critiques and engaging UN-Habitat in debate, but also for adapting global (urban) development policies and playing a role in their implementation, as we discuss next.

The Role of Higher Education In Relation to SDG and NUA Implementation

Higher education institutions, with urban geography, built environment, planning, urban studies and related departments and centres, generate knowledge through critical research about the challenges cities face and how policy responses work (Schorr et al. 2021). They contribute to the development of solutions, increasingly in dialogue with practitioners, communities and local governments; they train future practitioners and researchers needed to steer current transformation processes (*ibid.*). At first sight, one may gain the impression that universities are playing their role. Specialised degree programmes in sustainable urban development have been established in the last two decades, but mainly in institutions in the global North (Moore et al. 2015), formats for co-creating solutions between academia, policy and practice, such as Urban Living Labs, have become widespread (Baxter 2022), and the number of publications on sustainable urban development has skyrocketed (Sharifi 2021).

In the case of African universities, however, it appears that many are currently not adequately equipped to contribute to urban sustainability transitions, which understood as multi-stakeholder and cross-sectoral collaborative efforts to create new organisational structures and ways of doing and thinking that contribute to achieving sustainability (Frantzeskaki et al. 2017). A recent review of urban planning programmes across the continent shows that many do not adequately address the pressing challenges facing African cities, such as climate change, poverty and informality (Scholz et al. 2021). To explain the current state of planning education in African countries, the next section delves into the context of the colonial history of planning, which has been difficult to overcome.

From Modernist Colonial Planning to the Introduction of Urban Management (in Practice and in Academia)

The practice of urban planning in most countries has its origins in the desire of colonial powers to control and steer the development of settlements, particularly with regard to the health of colonial settlers and protection against the spread of infectious diseases (Odendaal et al. 2015). In many formerly colonised countries on the African continent, planning regulations were transferred from colonial to post-colonial centres and still form the basis of current planning systems, some of which were not revised until well after independence (Odendaal 2012). Today, planning systems, ideals and legislation across Africa still show strong continuities with the colonial era (Watson 2003; Scholz et al. 2021; Inkoom 2018). This can be seen at different levels. Colonial administrations in many African territories planned neighbourhoods for different segments of the population with very different densities and highly unequal levels of service provision, favouring the needs of European settlers and elites (Watson 2009). The idea of colonial cities as 'domains of European civility' (Odendaal et al. 2015, p. 287) is also reflected in the application of Western planning ideals, such as the application of garden city principles in Lusaka (Zambia) and Nairobi (Kenya) *(ibid.)* and tools such as master plans. The latter, a legacy of modernist planning, are often modelled on inappropriate visions of utopian urban and western futures, are unrealistically positivist and technocentric, and embody rational ambitions. They have long been subject to rigorous scholarly critique (see Mabogunje 1990; Diaw et al. 2002; Satgé and Watson 2018). As Sihlongonyane (2015, p. 8) contributes to this critique by arguing that "planning is constructed out of the normative values and sensibilities of the English-speaking dominant class"; modernist planning, at its core, disregards non-western ways of being urban, with a result that "[b]lack subjectivities are a misfit in the city as scenes of informality, black African practices, e.g. ritual cow slaughter, traditional healing shrines, customary circumcision shrines, burial systems and ancestral rituals convey expressions of dysfunctionality".

Invariably, the modernist planning approaches inherited by post-colonial African nations are unsuited for the complexity of African cities, contributing to the negative consequences of rapid urbanization, to the alarming increase in urban poverty, inadequate access to housing and basic urban services and the degradation of the urban environment. Although the rise of urbanisation challenges on the continent have long rendered modernist planning interventions ineffective, these structures and ideas still shape the contemporary urban form of the cities and continue to influence current developments.

Higher education systems in many African countries, like the town planning systems, were established by the colonisers. As such, they too are strongly embedded in the academic systems and traditions of the former colonial centres, including in terms of the dominance of Northern/Western epistemologies, theories and modes of teaching and research (Adriansen and Madsen 2019). The curricula of urban planning programmes on the African continent that were mostly established after independence were, in line with the planning systems, strongly influenced by those of the former colonial powers and the educators trained in the universities or educational systems of colonial powers (Watson and Odendaal 2012). Thus the programme established at Kumasi College of Arts, Science and Technology in Ghana in 1958 was designed to prepare students for the examination of the Town Planning Institute of Great Britain (Diaw et al. 2002). As Inkoom (2018, p. 145) puts it, urban planning curricula on the African continent "have not moved away from the […] curricula inherited from the colonial past".

In the resistance to reform, planning curricula have been rendered inappropriate and ineffective in the context of contemporary African cities. They obscure the role of people's practices of socio-economic organising and sustaining their livelihoods including efforts to generate income in the absence of wage labour, and to access and provide infrastructural services and shelter and how these shape urban life, politics and materiality (Wesely and Allen 2019). The need in African planning and architecture schools for "pedagogy to address, maintain, and increase local agency in the designing and building of its environments […]" is therefore plainly evident (Mukasa 2021, p. 132). They need to address "the blind spots reinforced by outdated colonial curricula" (Wesely and Allen 2019, p. 140).

Though perhaps initially evading the difficult task of fundamentally reforming the entrenched urban planning system and its curricula within universities, an understanding that a focus on 'urban management' may more effectively mitigate the challenges of urban growth arose out of the failure of planning. Since the mid-1980s, urban management has gained increasing importance, coinciding not only with rapid urbanisation but also with a wave of decentralization programmes (Naidoo et al. 2021:36). Under an urban management approach, the focus at all government levels has been beyond the spatial fix of planning to look at, firstly, ways to achieve better value for capital expenditures in providing urban services, utilities, land and housing; secondly, how to cut wastage through removing inefficiencies and 'leakages'; thirdly, how to improve cost recovery; and lastly, how to ensure a greater level of financial replicability for such programmes, so that these are ongoing, and do not come to an end once the budgets runs out (Jones and Ward 1994:33). According to the dedicated *Journal of Urban Management* (n.d.), urban management has emerged as a discipline concerned with an integrated approach to "planning, administering, regulating, and governing (PARG) urban complexity" (unpaginated). However, key to urban management, and in parallel to this, urban management pedagogy, is data and knowledge production, to which we turn next.

From Inequalities in African Urban Knowledge Production to North-South Cooperation (with Limitations)

In a reflection on gaps in its urban planning curriculum at the University of the Witwatersrand's planning programme around 2010, lecturers identified the unavailability of appropriate urban data and the inaccessibility of publications on African urbanism (Mabin et al. 2010). As co-authors of this chapter based at that university recall, urban management had also surfaced in curriculum discussions at this time, paralleled by discussions, for instance, at the University of Nairobi. The discussion about knowledge production on urban Africa and the challenge of limited data availability (Robin et al. 2019) is as relevant to planning as to the nascent initiatives to establish curricula at African universities.

Dominance and bias in academic publication is beginning to be scrutinised. It is becoming recognised that urban development in African countries is strongly underrepresented in academic publications in the fields of urban studies and planning (at least those listed in the Web of Science) (Sharifi et al. 2023). Few African universities are represented in the respective global academic discourse, South African higher education institutions being an exception (Walker and Boamah 2019). Those who publish on Africa in peer reviewed journals are predominately not based at African institutions (Crawford et al. 2021; Kemajou et al. 2021). A recent study on English and French language publications focussing on urban planning in Africa revealed that scholars from South African institutions published the largest number of articles in absolute numbers, while most publications are from researchers based at institutions in North America and Europe (Kemajou et al. 2021). West Africa, with the exception of Nigeria and Ghana, are particularly underrepresented when it comes to the locations studied and the authors' institutional affiliations (*ibid.*).

The reasons for these imbalances are manifold. They range from the underrepresentation of African scholars in editorial boards, which translate to biased attitudes towards research on Africa and from African institutions (Hedding and Breetzke 2021; Crawford et al. 2021), the chronic underfunding of many African universities, compounded by heavy teaching loads many scholars are faced with, to insufficient infrastructure and limited access to resources such as academic literature and research funding (Crawford et al. 2021; Kemajou et al. 2021; Adriansen 2020). As Northern/Western funding bodies are a main source of funding, African institutions also face challenges regarding their eligibility to apply for funds. Many global North funders require academics based on the African continent to 'partner' with institutions in the respective donor countries (Crawford et al. 2021), but they seldom consider the nature and implications of such partnerships, few being truly equal.

International cooperation in higher education between institutions in African countries and institutions in the global North has therefore played an important but ambivalent role in the development of research and higher education programmes in Africa. On the one hand, it allows for cross-learning between the partners involved (Oloruntoba 2021) and provides opportunities to tackle inequalities, for instance through funding, mentorship and training programmes for African scholars to publish in high ranked journals (Crawford et al. 2021). Global North-funded higher education collaboration involving scholars based at African universities may provide capacity building programmes at institutional level and even resources for curriculum reform, not only in line with sustainability agendas but also to Africanise educational programmes (Adriansen and Madsen 2019; Inkoom 2018). However, such North-South partnerships are embedded in the political, economic and epistemic inequalities of the global academic system and despite long standing critiques, often tend to reproduce those structures through dynamics such as the unequal control over funds, the prioritisation of specific research topics and of Western knowledge systems by funding bodies (Perry et al. 2023; Adriansen and Madsen 2019). Experiences from institutions located in other global South regions such as Latin America and South-East Asia mirror the challenges of establishing and shaping international cooperation with Northern universities as equals against the backdrop of imbalanced funding schemes as well as structures knowledge production and transfer (Leal et al. 2022; De Wit et al. 2017).

4 Conceptualising Curriculum Themes for Urban Management Within the Early Stages of the Wits-TUB Urban Lab

Aware of the dilemma of complex urban challenges and gaps regarding the training of future professionals and academics outlined above, the initial TUB-Wits collaboration under the SGD Graduate Schools initiative set out to create a new postgraduate programme in urban management at Wits with a transdisciplinary pedagogic approach. Located under a Masters of Urban Studies (MUS) with several fields, the offering was introduced in 2018 as the MUS in the field of Urban Management or MUS (UM), targeting urban practitioners across the African continent. The project team sought to leverage the insights of various networks for the development of key themes that could inform urban management curricula at Wits and beyond, on or for the African continent. Through the generous BMZ/DAAD funding, the project enabled the team to conceptualise and realise a series of workshops.

The Habitat III summit in October 2016, shortly after the inception of the first phase of Wits-TUB Urban Lab project, provided a unique opportunity to brainstorm the MUS (UM) curriculum approach and content. Through the initial project team's global networks of scholars and practitioners who happened to be at the conference, a highly interactive, unofficial side event could be organized. This sought to benchmark the urban management idea of the project and identify predominant urban management paradigms, thoughts and practices as well as innovative, creative ways of teaching the degree. Four themes crystallised in the preparation of this workshop and structured four break-away discussions:

- Politics and policy of the urban
- Understanding complex urban systems
- Managing change processes
- Co-producing knowledge between theory and practice.

Participants raised issues of transdisciplinarity, grassroot knowledge, complexity, collaboration and current technologies as significant. The notion of coproduction, the commons, New Urban Agenda were putatively in the air of the discussions. Participants based at higher education institutions felt the deliberations were useful for their own pedagogic endeavours.

A month later, Wits hosted the TUB counterparts and scholars from other African countries (Botswana, Kenya, Malawi) for a curriculum conference, which also engaged South African non-governmental organizations (NGOs), key officials from local government and the South African Local Government Association (SALGA). This further elaborated the four themes to develop an SDG orientated framework for the curriculum development at Wits. The enthusiastic engagement from experienced practitioners evidenced the significance of the curriculum initiative as responding to local realities. Participants stressed local collaboration, collaborative teaching and research as well as reciprocal benefit between the university and external collaborators.

A third joint curriculum workshop was hosted by TUB in the following year in July, alongside the first annual summer school of the project. This workshop engaged scholars and practitioners from Berlin and other German cities and included experts experienced in international cooperation, some based in funding institutions and others in NGOs, mostly working on urban issues in the global South. These deliberations, in a smaller round than the previous ones, generated an intense discussion, surfacing wicked problems such as climate change, complex human dynamics relating to gender and migration, and politics and economics of urban change.

The curriculum workshops and engagements facilitated the development of modules out of the four themes, which underpin the MUS (UM). To briefly elaborate, the aim of the first theme, *politics and policy of the urban*, is to prepare students for the challenging political environment, strengthen managerial and communicative competencies between built environment practitioners and political actors and provide an understanding of local government, political and planning processes in African cities. The second theme, *understanding complex urban systems*, provides competencies for work in largely 'informal' unpredicted developments as outlined in Sect. 2, and introduces innovative methods of spatializing data, holistic planning approaches and contextualization of urban problems to generate locally appropriate solutions. The third theme, *managing change process*, seeks to immerse students in a transdisciplinary environment, thus enhancing soft skills, improve conflict mediation, conduct stakeholder analysis and develop monitoring and evaluation systems. The aim of the fourth theme, *coproducing knowledge between theory and practice*, is to facilitate co-productive evidence-based and practice-oriented knowledge through fieldwork, collaborative projects and workshops, embed students within diverse institutions such as civil society, local government and the private sector and to develop urban studios and fieldwork as the key components of knowledge production.

5 Restructuring and Pedagogic Advances with the Introduction of an Urban Management Field at Wits

The cooperation between practitioners and educators in the field of urban management moves away from privileging technical knowledge alone in the production of 'integrated solutions' as demanded by the New Urban Agenda. It further transcends traditional hierarchisations and sectoralisation of knowledge and seeks to develop new formats for "co-producing" urban knowledge and policies. The outline of the core course of the MUS (UM) at Wits articulates the approach of degree as transdisciplinary, bringing together lecturers and practitioners into a practice of blended teaching and learning (Sihlongonyane 2019). Thus, the knowledge and expertise of residents, local initiatives, and civil society are brought into conversation with that of technicians and administrators, thereby responding to the above-mentioned debates on the need to restructure curricula at African universities. This also responds to the gap between planning and implementation, developing skills and capacities to enable the effective delivery of urban policies and programmes (*ibid.*).

The first task in the developing a master's degree in urban management in the School of Architecture and Planning at Wits was to consider the structure of the existing urban postgraduate offering (see Fig. 1) with a view to reforming this. Over several years, attempts at streamlining the School's urban masters degree had been met with resistance to change. This related to colleagues initiative over years having been invested in championing and sustaining specialised fields of study, as well as anxieties over the workload that restructuring involves, including onerous forms and approval rounds for curriculum development. As discussions were initiated with the academic convenors of the Masters in the Built Environment, MSc Development Planning, Master of Urban Design, Master of Urban Studies (MUS), Master of Architecture and Sustainable and Energy Efficient Cities, tensions were palpable. However, the above-mentioned workshops, in which some staff were able to participate, provided wider discussion and momentum moving the reform initiative forward. This was further sustained through standing item on the School Postgraduate Studies Committee agenda the monthly Planning Programme meetings. These provided space over several months for shifts in workloads and teaching duties to be discussed, revisited, negotiated and ultimately co-owned by all affected staff. Developing the motivation for the restructuring also involved liaising with the Wits School of Governance and the Wits School of the Social Sciences to establish potential overlaps or competition to the proposed field of urban management and to secure willingness for cross-disciplinary teaching.

The MUS (UM), the first urban management programme in the country and pioneering multidisciplinarity, was proposed and approved as a one year full-time or two years part-time degree. Offered as of 2018, the qualification navigates the challenge of teaching and producing regionally and locally relevant knowledge while keeping up with global discourses and standards. It seeks to educate and train planning or built environment practitioners, researchers, scholars and policy makers who can integrate the knowledge of and skills in urban policy, development and management and the planning, urban and social sciences in general in order to foster transformative urban strategies that enable cities to function more efficiently and effectively. Students are therefore attracted from different disciplinary backgrounds reflecting the diversity of disciplines and professional experiences relevant for the study and management of urbanisation in Africa.

The MUS (UM) curriculum is set out in Fig. 2. Compared to the urban management offering at TU Berlin and the one later introduced at UNILAG, the MUS (UM), as all masters coursework-and-research masters degrees at Wits, has relatively few courses. However, each course is substantial, covering a variety of topics. Some courses are shared in the school. 'Understanding Cities of the South' and 'Research Methods' are foundational across the MUS—since the restructuring, the MUS incorporates fields in Housing and Human Settlements, Urban Politics and Governance, Sustainable and Energy Efficient Cities and Urban Research, alongside Urban Management. The elective courses are offered for other fields and one is

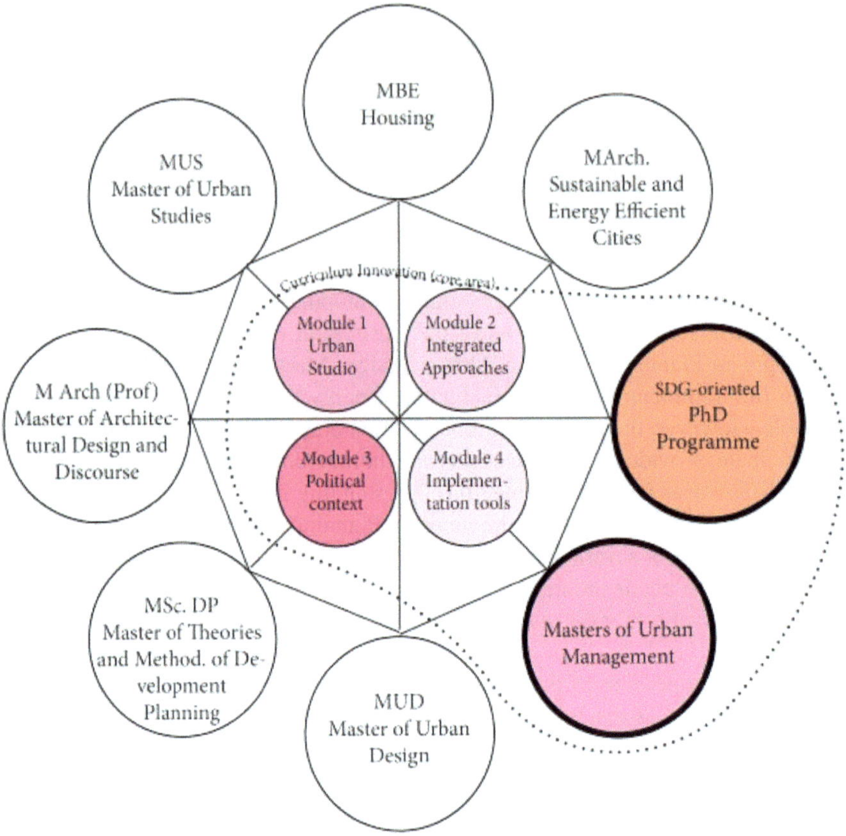

Fig. 1 The SDG Graduate School curriculum reform initiative with the Wits School of Architecture and Planning's masters and PhD offering in 2016 and the reform as implemented from 2019 (*source* Wits-TUB-UNILAG Uban Lab)

offered in Civil Engineering. Throughout, the curriculum encourages an interdisciplinary diagnosis/analysis and prognosis of urban issues across all scales. It provides graduates with an interdisciplinary theoretical, methodological, and practical base from which to approach the diverse issues involved in urban development. Therefore, it promotes an understanding of the complex urban processes and the development of effective interventions that can contribute to the development of well-managed sustainable, inclusive and resilient cities.

A range of teaching and learning techniques cut across these courses. These include seminars, lectures, discussions, guest lecturer inputs, field trips, reading assignments, presentations and computer applications. The core course, Theory and Practice of Urban Management, collaborates with the City of Johannesburg Metropolitan Municipality's Urban Management Unit as a core partner. This allows students to be exposed to actual, identified urban management challenges in the

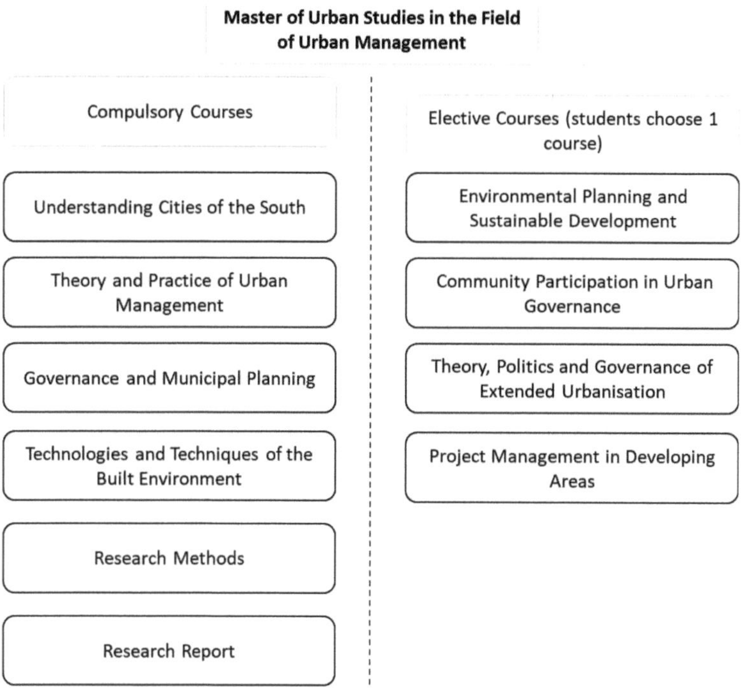

Fig. 2 Courses in the MUS (UM)

inner city. A recently added elective course, Theory, Politics and Governance of Extended Urbanisation, balances this with rapidly expanding peri-urban or semi-rural customary and neo-customary settings. Our late colleague and project coordinator, and co-author of this chapter Taki Sithagu's PhD research in one such area (see Sithagu 2022, 2024) formed a key case study and field trip destination for both Wits and University College London (UCL) students, as this course incorporates collaborative learning between masters Wits students and students from the Global Urbanism MASc programme in the Bartlett School of UCL. Much of this international collaborative teaching is made possible through online platforms such as Zoom, Teams, MOOC (for Massive Online Open Course) and Canvas, which the Covid-19 pandemic restrictions from 2020 to 2022 mainstreamed across higher education. The school has made significant upgrades to its Information Technology (IT) equipment, in part supported through the Wits-TUB-UNILAG Urban Lab, to enable collaboration across any distance.

The process of curriculum development for the MUS (UM) was a complex, bureaucratic and lengthy process. Curriculum restructuring and alignment of courses involved back-and-forth between course convenors, the chair of curriculum development unit, academic office and the project team at TUB. Issues of conflicting logics/ understanding of curriculum design in terms of integration, exclusivity, scale, detail, and framing of the courses were at the centre of many discussions. It became evident that the processes, procedures and timeframes for course approval differ considerably across higher education institutions.

The burst of the COVID 19 pandemic into the scene a few weeks into the 2020 MUS (UM) offering and at the cusp of Wits hosting TUB students for a summer school, caused initial disruption, cancellation and eventual recalibration with a new normal bedding itself in academic life. As elsewhere on the globe, new technology redefined teaching. Pedagogy has since had to respond to 'emerging technologies', which refers to a broad range of products and services that make use of artificial intelligence (AI), virtual reality (VR) and augmented reality (AR), wearable technology such as head-mounted displays and sensors, and the Internet of Things (IoT) made possible by 5G mobile standard (Almufarreh and Arshad 2023). Hundreds of digital education tools have been developed to increase student autonomy, enhance academic process management, promote collaboration, and improve communication between teachers and students. The use of e-learning tools and technologies has been empowering, but the adaptation has caused stress and anxiety. Whereas the COVID 19 experience enhanced capacity to communicate remotely, this seems to have compromised the human learning experience that comes with in-person engagement.

6 Long-Term Gains and Limitations of Introducing the MUS (UM) at Wits

The MUS (UM) curriculum was approved for implementation in late 2018. In the same year, TUB's urban management cohort participated in the Wits-hosted Summer School in Johannesburg in April. Twenty-six master's and PhD students from TUB joined 20 master's and PhD students from Wits. The purpose of the summer school was to test and pilot the four modules. The week-long summer school had teaching, a small studio and fieldwork. The studio was based in the Braamfontein Precinct as the lab for exploring various real-life challenges of urban change in the city of Johannesburg. The integrative elements involved combinations of lectures and fieldwork; the joint engagement of Wits and TUB students in projects; the engagement of partner organisations and lectures; as well as the consideration of different intertwining sectors e.g., housing, politics, and design in the Braamfontein area. Although, undertaken in a short space of time, the students managed to utilise various elements of implementation.

Although the participation of the Wits students was inconsistent, the summer school pilot project was an excellent way to enhance the programme while offering an opportunity for collaborative academic exchange over a week of various lectures, seminars, field trips, exhibitions and presentations by internal and external stakeholders. The students from Berlin expressed gratitude about the collaboration. One student, as a first-time visitor found the visit "an eye opener … as it dispelled some of the stereotypes about the continent." Another expressed how overwhelmed he was about the diversity of the country, but "some of the city was just like Europe." Some of the PhD and postdoctoral fellows funded through the project have since contributed to teaching in the School. The PhD students have also been central in the tutoring and mentoring of masters' students in the school, which in turn improved the throughput rate in the master's programme.

The availability of the WITS-TUB UNILAG-Urban Lab scholarships in the school offered by the project has attracted a attention to the school and the school has witnessed higher numbers of applications in the masters' degrees. At its first successful intake of UM students in February 2019, a total of 75 masters students (sub-Saharan African nationals and South Africans) between 2017 and 2023 were sponsored to study for one-year full time. Subsequently, the overall application numbers have increased from less than 100 to close to 200.

Therefore, the alumni footprint of the project is significant. It extends across the continent in Ghana, Nigeria, Kenya, Zimbabwe, South Africa, Uganda, Botswana, Namibia, Malawi, South Sudan, Ethiopia, and eSwatini. The scholarships have had the opportunity to open doors for citizens from sub-Saharan African countries other than South Africa to participate in the South African higher education system. They thereby have the opportunity to study in one of the best universities in the continent and become part of a "leading research-intensive university firmly embedded in the top 100 universities in the world…" (University of Witwatersrand n.d.).

The footprint could have been larger, had the intake not been stymied by the difficulty of securing visas and the increasingly complicated recognition of foreign qualifications in South Africa by the South Africa Qualification Authority (SAQA) amongst other challenges. The delays and decline of visa have undermined the thrust of the university to grow postgraduate numbers and be research intensive. The appointments of post-doctoral fellows from countries beyond South Africa also encountered delays to visa processes, in some cases necessitating cancellation or decline, which is a disappointing experience for the candidate as much as for the host.

Since the introduction of the MUS (UM) at Wits, the Habitat agenda has been noticeably engaged by staff in the programme signifying a move away from a South-African-centric focus. Both the research activities and the work on postgraduate curricula has enabled meaningful engagement with the SDGs. In particular, the project's work contributed to the operationalisation of Goal 11, while the research of the project's fellows and staff also provided starting points for formulating productive critiques of the universalist approach of the SDGs from an urban African perspective. Also, significant effort was made for each course in the MUS (UM) to meaningfully link with UN-Habitat's New Urban Agenda and the SDGs, especially SDG 11. For example, more course outlines across the Planning Programme in the School of Architecture and Planning at Wits now demonstrate a link with the Habitat agenda. Therefore, the project has contributed to breaking down the often South Africa-centric focus in teaching and to engendering more appreciation of the African perspectives and developments on the continent.

The collaborative teaching of *Theories and Practices of Urban Management* with officials of the City of Johannesburg has enhanced capacity to facilitate collaborative teaching. Within this collaborative spirit, the Wits-TUB-UNILAG Urban Lab has supported the co-hosting of conferences initiated in other research entities in the university.

The MUS (UM) must be self-sustained beyond the duration of the Wits-TUB-UNILAG Urban Lab and its bursaries. Increasingly, the degree relies on self-funded, part time students. To facilitate this, the project invested time and resources into a series of workshops from international to local, to devise a timetable that comfortable and affordably blends full time and part time teaching while also harnessing remote studying technologies. As with the curriculum restructuring, staff resistance to this long-envisaged initiative could be overcome through the structure that the project workshops and a project-funded timetabling expert consultant brought to this endeavour. Introduced in 2024, the new timetable is also contributing towards freeing time for research amongst staff.

The successful running of the UM programmes is attracting students from various parts of the continent. It has led to the establishment of networks across institutions in the continent through mobilities, exchanges and the conference it organised. Through hosting of webinars, talks and conferences hybrid teaching formats and hybrid forms of collaboration with other institutions, mainstreaming the format of webinars have been established. Through participation in conferences such as UN-Habitat's biannual World Urban Forum, and civil society networks such as Habitat International Coalition (HIC), the project teams' connections have extended beyond the continent. On an ongoing basis, scholars from other African Universities who Wits has hosted through the project, serve the School in capacities such as external examiners and reviewers, suggesting that lasting relationships have been built that also change the face of the otherwise South Africa-focussed School of Architecture and Planning.

7 Further Gains: The Urban Management Masters Programme at the University of Lagos

Nigeria's population crossed the urban majority threshold in 2015 (UN-Habitat 2015), with the urban patterns defying planning and management principles. It thus became imperative that local authorities and other actors at multi-level urban scales understand the changing patterns of urbanisation and can respond effectively to these changes through policy formulation and implementation within the context of sustainable development. However, there is a major gap in local governance capacities—especially regarding implementing the SDGs at the urban scale (Lawanson et al. 2021).

With only one university in Nigeria, Ahmadu Bello University, offering MSc and MPhil Urban management programmes in Nigeria, there was a significant gap in contemporary governance, policy and developmental studies in the country. This same gap is evident in the larger African context and global South in general as indicated above. In the lead-up to Habitat III, Policy Unit 4 on Urban Governance,

Capacity and Institutional Development highlighted a need for capacity development for public officials and other stakeholders at the municipal scale of development. This related to implementing the SDGs at the urban scale. Policy Unit 4 took its guidance from the requirements for the actualisation of the African Union Declaration (Agenda 2063), the New Urban Agenda of the United Nations (Habitat III), the Sustainable Development Goals, and the growing recognition of local authorities as an essential component of actualizing sustainable development (Habitat III, 2015). This represented a recognition of UNILAG's expertise in this area, while also allowing Professor Lawanson to plough insights from the Policy Unit into curriculum development at the University.

This, among other factors, led to the recognition in UNILAG for the need for a bespoke programme predominately targeted at local government officials. The Department of Urban and Regional Planning thus proposed a new interdisciplinary master's programme in Urban Management which was enthusiastically received. A committee was set up to write the programme. It leveraged on Taibat Lawanson's affiliation to erstwhile WITS-TUB Urban Lab and her time as visiting researcher at the University of Witwatersrand, which was developing its MUS (UM) at the time, and the Habitat Unit at TUB with its well-established urban management programme. These experiences were applied in the Postgraduate School at UNILAG with a curriculum proposal by 2018.

The imminent Master of Urban Management programme at UNILAG was a major rationale for UNILAG to join as a third partner to make up WITS-TUB-UNILAG Urban Lab from 2021 to 2025. The Masters in Urban Management programme was approved in 2021, alongside Master of Philosophy and PhD in Sustainable Urbanisation programmes. The urban management programme has since welcomed two cohorts of students from various disciplines. Admitted urban management students have bachelor's degrees in urban and regional planning, architecture, biochemistry, geography, mathematics, education and sociology. They also come from professional fields including filmmaking and photography, research administration, local government and civil service, as well as civil society and management consulting. Many of the students are recipients of scholarships and trilateral support under the WITS-TUB-UNILAG Urban Lab, including the week-long annual summer school hosting and collaborating with MUS (UM) students and the Wits-TUB-UNILAG Urban Lab project team in October 2022. Incidentally, the format of the summer school was so well received that other centres in UNILAG have adopted the model.

As with the MUS (UM) at Wits, the urban management curriculum at UNILAG was deliberately designed as an interdisciplinary programme aimed at training professionals who will be specialists in analysing and providing solutions for the policy formulation and implementation challenges of twenty-first century urbanisation, with a special focus on cities of the global South and the specific challenges these are facing. The objectives of the programme are to prepare students to assume leadership roles in government, non-profit organizations, and businesses by acquiring skills in five key areas:

- Mastery of knowledge in the history and theory of urban and regional development, the structure and functions of cities and urban systems, and local, national and global policy-making processes as it influences the fabric of human settlements in the global South, and especially in Nigeria.
- Proficiency in quantitative and qualitative research skills and their application to theory-building, data-gathering and analysis, and policy-making processes and infrastructure development in urban management and development contexts.
- Proficiency of communication skills for public (community) and policy settings.
- Awareness of the political, social, and ethical issues inherent in urban policy work and development practice.
- Capacity to formulate and implement policy solutions aimed at achieving global sustainable development benchmarks.

Given the diverse backgrounds of the students and the complex nature of urban challenges as set out earlier in this chapter, the curriculum is broad based, offering courses aimed at establishing the philosophy of urban management and cities, understanding the nature of cities, analysing the challenges of cities, and envisioning the future of cities (University of Lagos 2021). These range from globalization and regional development, development economics, municipal finance, urbanisation in Africa, project management to law and ethics, statistical methods for policy research and community advocacy seminars and a research dissertation (see Fig. 3).

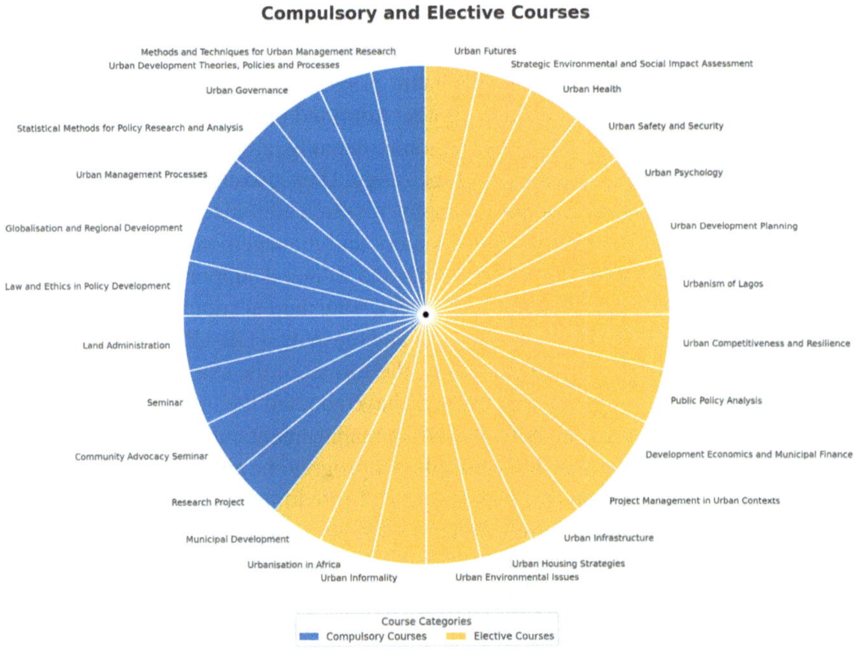

Fig. 3 Compulsory and elective courses in the Master of Urban Management at the University of Lagos

The programme has targeted working students which ensures immediate interaction with urban management and planning practice in Nigeria. Its part time structure combines contact lecture hours with graduate seminars in urban studies, governance, public policy and hands-on experience in community-based research, policy design, analysis, and implementation over a period of four semesters. Like the MUS (UM), teaching methodologies include problem-based learning that encompasses real-time urban challenges as learning scenarios, collaborative problem-solving exercises, and simulation exercises to enable students practice decision-making in complex, dynamic urban environments. Guest lectures from practitioners are also embedded into most courses in order to garner insights from professionals and practitioners in urban management and industry. Technology is deployed to have government officials and even UN-Habitat staff contribute to the teaching via zoom. Field visits are organised for on-site exploration of urban projects and interaction with stakeholders and practitioners and students are engaged with public governance discourse through such forums as the Lagos Resilience Forum, Physical Planning Summit etc. Other aspect of students' exposure include hybrid learning through virtual classrooms, discussion fora, webinars, and collaborative programmes in conjunction with the other partners.

The programme has been quite successful in that it attracted candidates from across Nigeria and from different disciplines, unlike other post-graduate programmes that attract candidates primarily from the Lagos and South West Nigeria axis. Students' projects covered a variety of issues from place making to water governance, community resilience, climate change and food security. With a number of WITS-TUB-UNILAG Urban Lab scholarships, 20 masters students and four PhD have been supported, with funded students participating in conferences, summer schools and research stays in Africa and beyond. The programme recently graduated the first set of students and there has been positive feedback. For example, one of the candidates from the first cohort said, "I did not want a regular masters degree of just attending lectures and writing exams. The MUM programme was interactive and challenging at the same time".

8 Early Career Researchers Advancing the Knowledge on African Urbanism

The urban management master offerings seek to build capacity across the continent for the urban challenges set out in the second and third section of the paper. Complimenting this, the Wits-TUB-UNILAG Urban Lab also seeks to impact on the aforementioned unevenness and limitations in knowledge production on the African continent by empowering earlier career academics. The inclusion of postdoctoral fellows both at Wits and at UNILAG ensured not only the generation of research additional to the work of the sponsored PhD students but also an agenda of early career researcher development.

The postdoctoral fellow team across TUB, Wits and UNILAG, while receiving their own career development support through the project, were tasked with develop managing and hosting a series of webinars and workshops across the three institutions targeting their peers, early career researchers, as widely across the African continent as possible. The webinars addressed and sought to dismantle the barriers to research within academic careers as well as to academic writing and publishing. In compiling the webinar programme, whose online participants regularly numbered in the order of 70, the postdoc team tapped into existing networks of the Wits-TUB-UNILAG Urban Lab but also served to widen this. The postdocs across the three institutions also collaborated with one another, initiating joint research papers and other collaborative endeavours. Their exposure through the Wits-TUB-UNILAG Urban Lab to debates on the global agendas at conferences of UN-Habitat and related bodies allowed for critical learnings and insights as well as sharing of knowledge and experience at the global level.

By the end of 2024, research by alumni and fellows had contributed to the above-mentioned co-produced book, in particular generating situated knowledge that contributes to a deeper understanding of the challenges facing African cities. This includes work exploring the transnational histories of the postcolonial built environment with a focus on Lagos (Oguntade 2025a, b; Harrison et al. 2024), and a particularly strong body of scholarship addressing the aforementioned complex of land management at the intersection of formal planning, customary land rights and informal forms of governance. This is a critical issue, especially in peri-urban areas of Africa experiencing rapid urbanisation, where customary land governance systems collide with formal planning. The project's PhD research explores these dynamics and potential policy responses to the resulting conflicts in Ghana and Eswathini (Sumbo 2022; Sumbo et al. 2023). In the case of Lagos, work by one of the post-doctoral researchers shows how the formalisation of land governance and planning contributes to the dynamics of socio-spatial exclusion in processes of urban expansion (Oyalowo 2022a). Other research by project members examines the functioning of hybrid land governance systems resulting from the coexistence of statutory and customary land rights in South Africa (Sithagu 2022, 2024) and the impact of these hybrid systems on women's access to land in Sierra Leone (Turay et al. 2024; Turay and Omirin 2023).

Drawing on diverse experiences of urbanisation across the African continent, project members engaged in the design and evaluation of planning strategies and tools aimed at developing more equitable and sustainable urban areas (Adeyemi et al. 2023; Rukambe and Irurah 2021). The team also critically reflected on international sustainable development norms and policies (Sihlongonyane 2025), and proposed approaches to sustainable urban development for a post-2030 agenda that are grounded in African realities, knowledge systems and everyday urban practices (Appelhans et al. 2025; Okunola et al. 2023).

In line with the projects' objectives, some of the alumni and postdoctoral work is evolving around the role of higher education systems for sustainable development, exploring the intersections between research agendas and universities' potential service to communities and society (Oyalowo 2022b), promoting equity in higher education through digital tools and platforms (Anane et al. 2022; Anane and Adusei 2024), and developing innovative pedagogical approaches for the urban disciplines (Oyalowo et al. 2022).

The publications include both articles and chapters published in international journals and with international publishers but also contributions to national discourses as well as non-academic publications such as blog posts that increase the visibility of the research (e.g. Turay 2023; Gondwe 2023; Oluwase 2024). Providing support for open access to journal articles through the project has facilitated the citation record of researchers in the project.

Commitment to collaboratively editing the book (Appelhans et al. 2025) mentioned in the introduction above and hosting the African Urbanisms conference at Wits (October 2024) further elevated relevant research that would embed the urban management degree initiatives within new knowledge generation. In addition to providing important outlets for the research of early career researchers from African institutions, working on a joint book and a scientific conference was also of great relevance for the academic staff involved in on capacity building within this project. These types of activities allow engagement with the project to be linked to quantifiable scholarly outputs, which are crucial for individual scholars and their career development in the prevailing 'publish or perish' systems, where engagement capacity building tends to receive little recognition from institutions. This contradiction is rarely considered by academic institutions and funding bodies.

9 Reflection and Conclusion

There is no doubt that the Wits-TUB and later Wits-TUB-UNILAG Urban Lab project made a positive impact in enabling the collaborative creation of two urban management programmes on the African continent, capacitating (future) practitioners and supporting the work of early career researchers and thereby contributing to creating and exchanging knowledge on urban development processes and dynamics across the African continent. Nevertheless, the project required navigating and bringing together different needs, requirements and interests.

The Urban Management Masters programmes at Wits and UNILAG both contribute to filling the gap in planning programmes that adequately address the challenges facing urban Africa. Both programmes take local and regional needs as their starting point and represent attempts to offer educational programmes that balance the need to address urban development challenges specific to their regional contexts by embracing local forms of knowledge (as discussed in Sect. 2), while at the same time providing postgraduate training that meets international standards and speaks to global discourses. Embedding the programmes in international exchanges with faculty and students from other African countries and Germany helped to broaden the perspectives of the students and scholars involved. The SDGs, as a common thread, provide valuable links to broader African and global discourses on urban development. The reference to the SDGs was also helpful in the development and establishment of programmes that focus on urban development and are institutionally embedded in planning schools but incorporate transdisciplinary approaches in their curricula and pedagogy.

Embarking on curriculum planning for new degrees and seeing this through to implementation requires long-term commitment from the involved institutions and staff. However, the demands placed on academics in terms of research outputs, teaching, administrative tasks and job mobility complicate such a commitment. This landscape is rife with workload discord between research time and time for teaching and doing administrative tasks. The design and implementation of capacity-building programmes and projects must therefore take account of these needs and circumstances to promote capacity development on the part of project staff.

Funding PhD and post-doctoral research, organising supportive webinars and workshops for early career researchers, and facilitating conference participation and exchanges has enabled postgraduate training to be closely linked to the production of knowledge on the dynamics of urban development in Africa. While these activities have had an impact and contributed to broadening understanding of the challenges facing urban Africa, many of the resulting publications are inaccessible to most African academic institutions, as access to and open access publication in international journals remains extremely costly, an ongoing challenge beyond the scope of the project. More debate and discussion need to be garnered to facilitate equitable knowledge production by promoting local knowledges, indigenous knowledge and innovative knowledge systems from the continent.

The international collaboration and funding opportunities provided important impetus to catalyse and enhance ongoing debates and developments in curriculum development at Wits and UNILAG and allowed for implementation to be supported through scholarships. Tapping into ongoing debates and collectively developing the funding application with all partners by utilising the scope of the project call allowed to largely match with the thematic and strategic agendas of DAAD and the partnering universities avoiding some of the pitfalls of North-South collaborations discussed above. Some inequalities, such as the global North institution being the key partner for the donor with implications for communication and financial management were reproduced as they are inscribed in the logics of allocating public funds in Germany. Further engagement to reckon with diverse methodological approaches, contextual intricacies and nuanced context considerations could contribute to mitigate inherent hierarchies and technologies of power to leverage the success of projects of this nature. In sum, the Wits-TUB-UNILAG Urban Lab experiences show that if the different needs and interests of participating institutions and academics can be balanced or even matched; multiple benefits can flow from such North-South cooperation projects, most importantly impact on human capacity building and on complex challenges such as the African urban reality which this project sought to engage with in meaningful ways.

References

Abbey M (2020) Addressing the challenges of urbanization in Africa: a summary of the 2019 African Policy Circle discussions. African Policy Circle. Position paper, Konrad Adenauer Foundation and Global Public Policy Institute. Available at: https://www.kas.de/documents/252038/7995358/Addressing+the+Challenges+of+Urbanization+in+Africa.pdf/df4e7f62-c130-e702-9669-0a746596028e. Accessed 17 Nov 2023

Adeyemi SA, Onifade VA, Jimoh HO, Akindeju F (2023) Development control as a tool for urban sustainability in south-western Nigeria. Urban Reg Plan Rev 9(1):10–24

Adriansen H, Madsen L (2019) Capacity-building projects in African higher education: issues of coloniality in international academic collaboration. Learn Teach 12(2):1–23

Adriansen H (2020) Materialities and mobilities in transnational capacity building projects: uneven geographies of knowledge production. Popul Space Place 26(3):1–12

AfDB (2012) Urbanization in Africa. African Development Bank Group (AfDB) Blog. Available at: https://blogs.afdb.org/inclusive-growth/urbanization-africa-191. Accessed 15 Nov 2023

Agheyisi EJ (2012) Channels and institutions of informal land delivery in the urbanizing fringes of Benin City. Sokoto J Soc Sci 2(1):73–88

Akaateba M, Huang H, Adumpo E (2018) Between co-production and institutional hybridity in land delivery: insights from local planning practice in peri-urban Tamale, Ghana. Land Use Policy 72:215–226

Almufarreh A, Arshad M (2023) Promising emerging technologies for teaching and learning: recent developments and future challenges. Sustainability 15(8):1–21

Anane GK, Adusei A (2024) Democratizing online learning platforms: Ghanaian universities' experience in promoting equity through accessibility in online learning. In Kyei-Blankson L, Keengwe J, Ntuli E (eds) Designing equitable and accessible online learning environments (democratizing-online-learning-platforms). IGI Global

Anane GK, Addo PK, Adusei A, Marfo C (2022) Risk management strategies for teaching and learning in Ghanaian public universities during the COVID-19 pandemic. In: Namatende-Sakwa L, Lewinger S, Langsford C (eds) COVID-19 and education in Africa: challenges, possibilities, and opportunities. Taylor & Francis

Appelhans N, Rawahani C, Huchzermeyer M, Oyalowo B, Sihlongonyane MF (eds) (2025) Everyday urban practices in Africa: disrupting global norms. Routledge, Oxon

Aust H, du Plessis A, French D, Kotzé LJ (2018) Good urban governance as a global aspiration: on the potential and limits of sustainable development goal 11. In: Sustainable development goals. Law, theory and implementation. Edward Elgar Publishing, pp 201–221

Baxter J (2022) From urban living labs towards critical spatial design: decolonising knowledge in urban design and planning using critical reflexivity. Nordic J Urban Stud 2(2):124–140

Brouwer R, Nhassengo J (2006) About bridges and bonds: community responses to the 2000 floods in Mabalane district, Mozambique about bridges and bonds. Disasters 30(2):234–255

Commins S (2018) From urban fragility to urban stability. Africa Security Brief, No. 35, 12 June. Available at: https://africacenter.org/wp-content/uploads/2018/07/ASB35EN-From-Urban-Fragility-to-Urban-Stability.pdf. Accessed 14 Nov 2023

Crawford G, Mai-Bornu Z, Landström K (2021) Decolonising knowledge production on Africa: why it's still necessary and what can be done. J Br Acad 9s1:21–46

Croese S (2022) Toward an embedded, integrated, and collaborative approach to SDG localization in African cities. In: Croese S, Parnell S (eds) Localizing the SDGs in African cities. Springer, Cham, pp 1–11

de Satgé R, Watson V (2018) Urban planning in the global south: conflicting rationalities in contested urban space. Macmillan, Palgrave

De Wit H, Gacel-Ávila J, Jones E (2017) Voices and perspectives on internationalization from the emerging and developing world, where are we heading? In: De Wit H, Gacel-Ávila J, Jones E, Jooste N (eds) The globalization of internationalization. Routledge, London, pp 221–233

Diaw K, Nnkya T, Watson V (2002) Planning education in sub-Saharan Africa: 'responding to the demands of a changing context.' Plan Pract Res 17:337–348
Diaz-Sarachaga JM, Sarachaga JL (2024) Diagnosis on the implementation of the New Urban Agenda. Glob Sustain 7(e37):1–7
Durand-Lasserve A (2004) Land for housing the poor in African cities: are neo-customary processes an effective alternative to formal systems? (Report). Centre National de la Recherche Scientifique, Université Denis Diderot, Paris
Eisenmenger N, Pichler M, Krenmayr N, Noll D, Plank B, Schalmann E, Wandl M-T, Gingrich S (2020) The Sustainable Development Goals prioritize economic growth over sustainable resource use: a critical reflection on the SDGs from a socio-ecological perspective. Sustain Sci 15(4):1101–1110
Eriksson A (2007) Special report: silent disasters. Nurs Health Sci 9(4):243–245
Frantzeskaki N, Broto V, Coenen L, Loorbach D (2017) Urban sustainability transitions. Routledge, London
Gondwe J (2023) Bringing animal geographies to the ECAS 2023 conference. Available at: https://www.urban-sdg-school.org/blog/post/ECAS-2023. Accessed 8 Dec 2023
Güneralp B, Lwasa S, Masundire H, Parnell S, Seto K (2018) Urbanization in Africa: challenges and uncertainties and implications for opportunities for conservation. Environ Res Lett 13(015002):1–8
Harrison E, Jackson I, Appeaning Addo I, Muraina O (2024) "Kingsway leads the way to modern living": British Profit-seeking and Modernism in Ghana and Nigeria 1920–1970. J Des Hist 37:217–235. https://doi.org/10.1093/jdh/epae010
Hedding DW, Breetzke G (2021) "Here Be Dragons!" The gross under-representation of the global south on editorial boards in geography. Geogr J 187(4):331–345
Homan J (2003) The social construction of natural disaster: Egypt and the UK. In: Pelling M (ed) Natural disaster and development in a globalizing world. Routledge, London, pp 141–156
Habitat III (2015) Urban governance, capacity and institutional development. Policy Unit 4. United Nations Conference on Housing and Sustainable Urban Development, Nairobi
Huchzermeyer M, Appelhans N, Oyalowo B, Sihlongonyane M (2025) Global norms, urban Africa, the everyday and disruption. In: Appelhans N, Rawhani C, Huchzermeyer M, Oyalowo M, Sihlongonyane MF (eds) Everyday urban practices in Africa: disrupting global norms. Routledge, London
Huq S, Kovats S, Reid H, Satterthwaite D (2007) Editorial: Reducing risks to cities from disasters and climate change. Environ Urban 19(1):3–15
Ikejiofor U (2009) Planning within the context of informality: issues and trends in land delivery in Enugu, Nigeria. Case study prepared for Revisiting Urban Planning: Global Report on Human Settlements 2009. Available at: https://citeseerx.ist.psu.edu/document?repid=rep1&type=pdf&doi=75b4141cab82007fe5860641004db5f512361bd7. Accessed 17 Nov 2023
Inkoom D (2018) Getting unstruck: planning education legacies and urban challenges in the Global South. In Grubbauer M, Shaw K (eds) Across theory and practice: thinking through urban research. Jovis, Berlin, pp 140–148
Jones G, Ward P (1994) The World Bank's 'new' urban management programme: paradigm shift or policy continuity? Habitat Int 18(3):33–51
Journal of Urban Management (n.d.) Aims and scope—Journal of Urban Management. Journal website. Available at: https://www.sciencedirect.com/journal/journal-of-urban-management/about/aims-and-scope. Accessed 19 Nov 2023
Kaika M (2017) 'Don't call me resilient again!': The New Urban Agenda as immunology … or … what happens when communities refuse to be vaccinated with 'smart cities' and indicators. Environ Urban 29(1):89–102
Kemajou A, Konou AA, Jaligot R, Chenal J (2021) Analyzing four decades of literature on urban planning studies in Africa (1980–2020). Afr Geogr Rev 40(4):425–443
Kombe W (2005) Land use dynamic in peri-urban areas and their implications on the urban growth and form: the case of Dar es Salaam, Tanzania. Habitat Int 29(1):113–135

Lawanson T, Oyalowo B, Nubi T (2021) Delivering the global urban development agenda in Lagos, Nigeria—a Local lens is needed. In: Nubi T, Anderson A, Lawanson T, Oyalowo B (eds) Housing and SDGs in Africa. Springer, Singapore
Leal F, Finardi K, Abba J (2022) Challenges for an internationalization of higher education from and for the global south. Perspect Educ 40(3):241–250
Liu Z, Eden J, Dieppois B, Conradie W, Blackett M (2023) The April 2021 Cape Town wildfire: has anthropogenic climate change altered the likelihood of extreme fire weather? Bull Am Meteor Soc 104(1):299–304
Mabin A, Todes A, Boshoff B, Benit-Gbaffou C, Karam A, Klein G, Klug N, Sihlongonyane M, Winkler T (2010) Through the looking glass: planning education at the University of the Witwatersrand, Johannesburg. J Build Land Dev 17(3):156–173
Mabogunje A (1990) Urban planning and the post-colonial state in Africa: a research overview. Afr Stud Rev 33:121–203
Moore S, Rydin Y, Garcia B (2015) Sustainable city education: the pedagogical challenge of mobile knowledge and situated learning. Area 47(2):141–149
Mudefi E (2023) Disaster management 'deeds' in the context of April 2022 KwaZulu-Natal floods: a scoping review. Int J Disaster Risk Reduct 98:1–21
Mukasa L (2021) Teaching African architecture: academic education and identity. In: Meuser P, A Dalbai (eds) Theorising architecture in sub-Saharan Africa. perspectives, questions and concepts. Dom Publishers
Nagati O, Gad H, El-Didi AA, Kihila JM, Mbuya E, Njavike E (2023) Towards a bottom-up approach for localising SDGs in African cities: findings from Cairo and Dar es Salaam. Afr Dev/Afrique Et Développement 48(1):79–112
Naidoo C, Nair J, Ngcobo L (2021) A critique of the modernist approach to post-apartheid housing delivery and urban design. J Inclus Cities Built Environ 1:25–37
Nur I (1999) Current extent of disasters in Africa. Prehosp Disaster Med 14(2):66–74
OECD/UN ECA/AfDB (2022) Africa's urbanisation dynamics 2022: the economic power of Africa's cities. Organisation for Economic Cooperation and Development, United Nations Economic Commission for Africa and the African Development Bank. Paris: OECD Publishing. Available at: https://www.oecd-ilibrary.org/development/africa-s-urbanisation-dynamics-2022_3834ed5b-en. Accessed 19 Nov 2023
Odendaal N (2012) Reality check: planning education in the African urban century. Cities 29(3):174–182
Odendaal N, Duminy J, Inkoom D (2015) The developmentalist origins and evaluation of planning education in sub-Saharan Africa, C. 1940 to 2010. In: Nunes Silva C (ed) Urban planning in sub-Saharan Africa. Colonial and post-colonial planning cultures. Routledge, New York, pp 285–299
Oguntade B (2025a) Bookshop House, Lagos 1973. In: International D (ed) Modernism in Africa: the architecture of Angola, Ghana, Mozambique, Nigeria, Rwanda, South Africa, Sudan, Tanzania, Uganda. Birkhäuser, pp 260–263
Oguntade B (2025b) Main library. University of Lagos (UNILAG), Lagos, 1964. In: Docomomo International (ed) Modernism in Africa: the architecture of Angola, Ghana, Mozambique, Nigeria, Rwanda, South Africa, Sudan, Tanzania, Uganda. Birkhäuser, pp 240–244
Okunola OH, Balogun K, Ojo DB (2023) Indigenous knowledge and flood resilience strategies in African coastal cities: from practice to policy. In: Shahzad N (ed) Water and environment for sustainability: case studies from developing countries. Springer, Cham, pp 161–172
Oloruntoba S (2021) The state and the state of knowledge production in African universities: rethinking identity and curricula. In: Afolayan A, Yacob-Haliso O, Oloruntoba SO (eds) Pathways to alternative epistemologies in Africa. Springer, Cham, pp 61–75
Oluwase (2024) How African cities can learn from each other about building climate resilience. World Econ Forum. Available at: https://www.weforum.org/stories/2024/04/african-cities-climate-change-resilience/. Accessed 10 Feb 2025

Oyalowo B (2022a) Implications of urban expansion: land, planning and housing in Lagos. Build Cities 3(1):692–708
Oyalowo B (2022b) An agenda for transdisciplinary research for achieving African higher education's third mission. J Educ Stud 21(1):11–29
Oyalowo B, Akinpelu D, Nubi T (2022) Reshaping housing pedagogy and public policy through documentaries in Lagos, Nigeria. Int J Hous Policy 22(3):395–413
Perry B, Castán Broto V, Patel Z, Sitas R (2023) Alliances, allyship and activism: the value of international partnerships for co-producing just cities. Plan Theory 23(4):423–445
Rakodi C (2007) Land for housing in African cities: are informal delivery systems institutionally robust and pro-poor? Glob Urban Dev 3(1):1–14
Ravallion M (2001) On the urbanization of poverty. Policy Research Working Paper No. 2586. The World Bank, New York
Robin E, Steenmans K, Acuto M (2019) Harnessing inclusive urban knowledge for the implementation of the new urban agenda. Urban Res Pract 12(2):137–155
Rukambe V, Irurah D (2021) Memories of futures—past and visions of future-futures: an architecture-to-backcasting metaphor approach towards tools and practices for sustainable city transitioning in Africa. In: Addaney M, Cobbinah P (eds) Sustainable urban futures in Africa. Routledge, Oxon
Scholz W, Stober T, Sassen H (2021) Are urban planning schools in the global south prepared for current challenges of climate change and disaster risks? Sustainability 13(3):1–16
Schorr B, Braig M, Fritz B, Schütt B (2021) The global knowledge value chain on sustainability: addressing fragmentations through international academic partnerships. Sustainability 13(17):1–20
Sharifi A (2021) Urban sustainability assessment: an overview and bibliometric analysis. Ecol Ind 121:1–18
Sharifi A, Khavarian-Garmsir AR, Allam Z, Asadzadeh A (2023) Progress and prospects in planning: a bibliometric review of literature in urban studies and regional and urban planning, 1956–2022. Prog Plan 173:1–37
Sihlongonyane MF (2015) Empty signifiers of transformation in participatory planning and the marginalization of black people in South Africa. Plan Pract Res 30(1):1–18
Sihlongonyane FM (2019) Theories and practices of urban management. Unpublished course outline. School of Architecture and Planning University of Witwatersrand, Johannesburg
Sihlongonyane MF (2025) Questioning the urban-centrism of the new urban agenda and its implications for African cities. In: Appelhans N, Rawhani C, Huchzermeyer M, Oyalowo B, Sihlongonyane MF (eds) Everyday urban practices in Africa. Routledge, Oxon, pp 25–39
Sithagu A (2022) The 'invisible' property system and revenue collection in former homelands in the context of hybrid governance and access to land and basic services. Transformation: Crit Perspect South Afr 110(1):76–92
Sithagu A (2024) The SDG monitoring framework turns a blind eye to the daily realities of lived tenure security in African hybrid land transaction systems: a south African case. In: Appelhans N, Rawhani C, Huchzermeyer M, Oyalowo B, Sihlongonyane MF (eds) Everyday urban practices in Africa. Routledge, Oxon, pp 172–188
Sulla V, Vikhali P, Cuevas F (2022) Inequality in South Africa: an assessment of the southern African customs union. The World Bank, Washington, DC. Available at: https://documents.worldbank.org/en/publication/documents-reports/documentdetail/099125303072236903/p1649270c02a1f06b0a3ae02e57eadd7a82. Accessed 15 Feb 2025
Sumbo DK (2022) Indigenes' exclusion from neo-customary land: a perspective from changes in usufruct rights in Pramso, Peri-Urban Kumasi—Ghana. Land Use Policy 120:1–12
Sumbo DK, Anane GK, Inkoom DKB (2023) 'Peri-urbanisation and loss of arable land': indigenes' farmland access challenges and adaptation strategies in Kumasi and Wa, Ghana. Land Use Policy 72:470–479

Technische Universität Berlin (TUB) (n.d.) Urban management, M.Sc. Technical University of Berlin website. Available at: https://www.tu.berlin/en/studying/study-programs/all-programs-offered/study-course/urban-management-m-sc. Accessed 29 Nov 2023

Tietjen B, Jacobsen K, Hollander J (2023) Climate change and urban migration in sub-Saharan African cities: impacts and governance challenges. J Clim Resilience Justice 1:20–32

TUB (n.d.) Urban management MSc.: programme overview. Technical University Berlin, Berlin. Available at: https://www.tu.berlin/en/studying/study-programs/all-programs-offered/study-course/urban-management-m-sc. Accessed 10 Feb 2025

Turay BT (2023) Women need access to land in Sierra Leone and New Laws don't go far enough. Africa at LSE. Available at: https://blogs.lse.ac.uk/africaatlse/2023/03/09/women-need-access-to-land-in-sierra-leone-and-new-laws-dont-go-far-enough/. Accessed 8 Dec 2023

Turay BT, Onifade VA, Jimoh HO, Akindeju IS (2024) The dynamics of access and exclusion of women to land resources. a case study of Tambakha Simibungie Chiefdom, North-West Province, Sierra Leone. Surv Rev 56(396):211–227

Turay BT, Omirin MM (2023) Improved land reforms to the benefit of women's access to land to foster and support social norm change, a case study of western area rural district and Bombali District in Sierra Leone. Afr J Land Policy Geospat Sci 6(3):400–417

UNCTAD (2018) Economic development in Africa report. Available at: https://unctad.org/en/PublicationChapters/edar2018_ch1_en.pdf. Accessed 29 Jan 2020

UNDRR (2008) Climate change and disaster risk reduction. Briefing note No 1. United Nations Office for Disaster Risk Reduction, Geneva. Available at: https://www.undrr.org/publication/briefing-note-01-climate-change-and-disaster-risk-reduction. Accessed 10 Feb 2025

UN-Habitat (2018) Tracking progress towards inclusive, safe, resilient and sustainable cities and human settlements. SDG 11 Synthesis Report—High Level Political Forum 2018. United Nations Human Settlements Programme, Nairobi. Available at: https://www.un-ilibrary.org/content/books/9789210472401. Accessed 15 Feb 2025

University of Lagos (2021) School of postgraduate studies, University of Lagos Handbook 2021–2023. University of Lagos, Lagos

University of Witwatersrand (n.d.) Master of urban studies in the field of urban management. Available at: https://www.wits.ac.za/urban-management/. Accessed 14 Oct 2020

Vives-Miró S (2022) The urbanization of poverty: rethinking the production of unjust geographies. Fennia: Int J Geogr 200(1):41–51

Walker M, Boamah E (2019) Making the invisible hyper-visible: knowledge production and the gendered power nexus in critical urban studies. Hum Geogr 12(2):36–50

Watson V (2003) Conflicting rationalities: implications for planning theory and ethics. Plan Theory Pract 4(4):395–407

Watson V (2009) 'The planned city sweeps the poor away…': Urban planning and 21st century urbanisation. Prog Plan 72:151–193. https://doi.org/10.1016/j.progress.2009.06.002

Watson V (2016) Locating planning in the new urban agenda of the urban sustainable development goal. Plan Theory 15(4):435–448

Watson V (2021) The return of the city-region in the new urban agenda: is this relevant in the Global South? Reg Stud 55(1):19–28

Watson V, Odendaal N (2012) Changing planning education in Africa: the role of the Association of African Planning Schools. J Plan Educ Res 33(1):96–107

Wesely J, Allen A (2019) De-colonising planning education? Exploring the geographies of urban planning education networks. Urban Plan 4(4):139–151

Wits (n.d.) Master of urban studies in the field of urban management. Wits University website. Available at: https://www.wits.ac.za/urban-management/. Accessed 29 Nov 2023

Wu H, Huang M, Tang Q, Kirschbaum D, Ward P (2016) Hydrometeorological hazards: monitoring, forecasting, risk assessment, and socioeconomic responses. Adv Meteorol 2016:1–3

RoHan Rostock-Hanoi DAAD SDG Graduate School: Catalysis as Key Towards Sustainable Resource Management

Dirk Hollmann, Esteban Mejía, Tabea Angela Thiel, Udo Kragl, Le Minh Thang, and Le Thanh Son

Abstract In this chapter, we intend to show how small-scale scientific collaborations in chemistry can develop into long-term strategic university partnerships, addressing the barriers, challenges, opportunities and impacts. Indeed, the RoHan DAAD SDG Graduate School (SDG—Sustainable Development goals) is regarded as a pioneering lighthouse project at the interface of science and international cooperation, aiming to build a bridge between research and development. Funded by the German Academic Exchange Service (DAAD), supported by funds from the German Federal Ministry for Economic Cooperation and Development (BMZ), research groups from Rostock and Hanoi have been connected to collaboratively work on innovative solutions in chemistry, particularly catalysis. This ten-year project is considered a unique opportunity to share knowledge through interdisciplinary and international cooperation and to develop promising technologies urgently needed in Germany and in developing regions such as Vietnam. The focus extends beyond doctoral students to include Master students, post-doctoral researchers, and faculty, underlining the holistic and integrative orientation of the program. By linking basic research with practical applications and international collaboration, RoHan is seen not only as an academic endeavour but as an example of commitment to research and education

D. Hollmann (✉) · U. Kragl
Interdisciplinary Faculty, Department Life, Light and Matter and Institute of Chemistry, University of Rostock, Rostock, Germany
e-mail: dirk.hollmann@uni-rostock.de

E. Mejía · T. A. Thiel
Leibniz Institute for Catalysis (LIKAT), Rostock, Germany
e-mail: Esteban.mejia@catalysis.de

L. M. Thang
Department of Chemical Engineering, School of Chemistry and Life Sciences, Hanoi University of Science and Technology, Hanoi, Vietnam
e-mail: thang.leminh@hust.edu.vn

L. T. Son
Faculty of Chemistry, VNU University of Science, Hanoi, Vietnam
e-mail: sonlt@vnu.edu.vn

C. Brandt et al. (eds.), *Implementing Sustainability and Internationalization in Higher Education*, World Sustainability Series, https://doi.org/10.1007/978-3-031-97837-1_6

that addresses global challenges and provides innovative solutions for a sustainable future.

1 Introduction

Within this chapter, we will first look at the relationship between Vietnam and Germany. This is one of the reasons for the success of this project. Minimally, we will go into the so-called topic Catalysis, which links the working groups in Chemistry. Here the author refers to the relevant specialised journals which provide specific information on this topic. This is not part of this book section. However, catalysis, a sub-area of chemistry, is very relevant to achieving the corresponding Sustainable Development Goals. It is therefore explicitly described and explained in context.

In the Analysis section we go into more detail of both RoHan phases. These are explicitly explained to give the reader an impression of the approach and the challenges from the start of funding to the end of funding. This can make it easier for the user to implement analogue strategies. The first part is about the beginning or the start of the project and holistic networking. The second part is mainly concerned with the realisation and establishment of a long-term network in order to ensure the continuation of the project even after the funding has ended. Of particular note here is the implementation of double-degree programmes. During the project, special attention was paid to the implementation of ICT methods for digitalisation. Furthermore, the so-called Leave No One Behind strategy was established for the students. We think that this is particularly relevant in the future and we included it.

We conclude with a summary and outlook on the impact that this project can have on strategic collaboration between individual universities and working groups. The RoHan project is a good example of how a bottom-up approach between scientists can lead to a long-term and comprehensive top-down approach between universities. We hope that the reader will find it easier to set up similar projects and that cooperation between developing and developed countries will be encouraged.

Scientific Relations Between Germany and Vietnam

The relations between Germany and Vietnam have evolved over time, with an increasing focus on education and research. Traditionally, Vietnamese universities were mainly financed by the Vietnamese government and therefore heavily dependent on state directives, leading to limited research capacities, inadequate infrastructure, and restricted international collaboration. The educational landscape in Vietnam was shaped primarily by state requirements, offering little flexibility and freedom for independent initiatives.

Between 2001 and 2015, Vietnam's international scientific collaboration with countries like the USA (e.g. Center for Molecular and NanoArchitectures—MANAR), Japan, South Korea, France, England, and Germany increased. This period saw growth in research output and heightened visibility for Vietnam-based research (Cordova and Yaghi 2019; Nguyen et al. 2017). However, a strong dependency on foreign collaborators persisted, as reflected in publication trends. Due to publication pressures—Master students receive only the best mark with a publication—95% of publications appeared in national Vietnamese journals in Vietnamese language, creating a scientific "bubble" and limiting global visibility.

Scientific and academic cooperation with Germany is particularly interesting for Vietnam for several reasons, especially due to its historical relationship with the former eastern part of Germany, so called German Democratic Republic (Deutsche Demokratische Republik, DDR). During the Cold War, close relations were established between the DDR and Vietnam, strengthened by an agreement to send Vietnamese workers and students to the DDR by filling labor shortages, particularly in manufacturing and construction, thereby contributing significantly to industrial productivity. Meanwhile, the Federal German Republic (Bundesrepuplik Deutschland, BRD) took in Vietnamese refugees and promoted their integration into society at great expense. This historical connection has had a lasting effect. Unlike the BRD, the DDR extended the social integration of contract workers and students into society, resulting in a strong Vietnamese visiting scholar community (between 42,000 and 50,000) (Bösch and Su 2018). Many of these Vietnamese visiting scholars, educated in the DDR, now hold leading positions in academia and politics in Vietnam. This generation of academics forms a strong bridge between the two countries and has an affinity with Germany and its culture. They are often advocates for the continuation and deepening of scientific cooperation between Germany and Vietnam. As an example, the Vietnamese-German University (VGU) in Binh Duong was founded in 2008, offering studies in engineering, natural sciences, and economics. The model of German higher education, with its principles of academic freedom, unity of teaching and research, and institutional autonomy, is followed by VGU.

Numerous other collaborations have been funded and supported by various German and Vietnamese foundations The German Federal Ministry of Education and Research (BMBF) has played an important role in promoting these collaborations, for example, through the Joint Funding Scheme for projects in fields like water management. The Higher Education Compass of the German Rectors' Conference (HRK, www.hochschulkompass.de) lists 188 official cooperation projects between Germany and Vietnam in 2024, involving 100 German universities cooperating with 59 Vietnamese universities and 4 other institutions. Scientific cooperation with Vietnam has been promoted by the German Research Foundation (DFG), which has advised the National Foundation for Science and Technology Development (NAFOSTED) in Vietnam on its establishment (DFG-NAFOSTED Cooperation 2024). Furthermore 2008 the Vietnamese-German University (www.vgu.edu.vn) was initialized in Ho-Chi-Minh-City. This university is based on a cooperation agreement between Vietnam and the German federal state Hessen. Diverse institutions collaborated with Vietnamese partners in various scientific fields have such as the Max Planck Society (MPG), the Leibniz Association, the Helmholtz Association of German Research Centers (HGF), and the Fraunhofer Society (FhG). Additionally, educational cooperation between the two countries has been strengthened by the Federal Institute for Vocational Education and Training (BIBB), which has supported the Vietnamese Institute for Vocational Training (NIVT) in various priority areas. The DAAD

Competence Centre for International Academic Cooperation (KiWi) is currently providing German universities and institutes with a cooperation guide for Vietnam (https://www.daad.de/de/infos-services-fuer-hochschulen/kompetenzzentrum/).

In 2016 the DAAD enabeled the established so called Sustainable Development Goals Graduate schools to enhance german university collaboration with developing countries. The RoHan Rostock-Hanoi DAAD SDG Graduate School: Catalysis as Key Towards Sustainable Resource Management (RoHan) project connects two local networks: the University of Rostock (UR) and the Leibniz Institute for Catalysis (LIKAT) in Rostock, Germany, and the Hanoi University of Science and Technology (HUST) and the Vietnam National University, University of Science, Hanoi (VNU-HUS) in Hanoi, Vietnam. Given their overlapping expertise in sustainable chemistry and catalysis, the partner institutions were well-suited for intensive collaboration. However, as is typical in science, these partnerships have largely been formed through individual projects. Before 2016, only small-scale collaborations and workshops have been conducted between the University of Rostock and VNU-HUS, as well as between LIKAT and HUST. These exchanges were often initiated through personal connections and supported by small budgets from the DFG and DAAD which always stopped after the funding.

At the University of Rostock, two faculties are involved: the Faculty of Mathematics and Natural Sciences (FMN) and the Faculty of Interdisciplinary Research, Department of Life, Light & Matter (LLM). The university's chemistry department provides extensive expertise in homogeneous and heterogeneous catalysis, biocatalysis, and organic chemistry, with a focus on organic synthesis, and biomass conversion. The Leibniz Institute for Catalysis (LIKAT) at the University of Rostock is Europe's largest publicly funded research institute focused on applied catalysis. Its main objectives are to acquire new knowledge in catalysis and apply it in real-world settings, particularly for sustainable processes. LIKAT's research includes developing innovative catalysis methods and technologies, with an emphasis on the sustainable use of resources. The Faculty of Interdisciplinary Research (LLM) has facilitated cross-linking between the FMN, LIKAT, and other institutes.

HUST, one of the largest and top-ranked technology universities in Vietnam, excels in catalysis research and innovation, ranking first in Vietnam in the QS World Ranking 2022 and Emerging Economies University Rankings 2021. Similarly, VNU-HUS, Vietnam's leading institution for basic science, has a long tradition in catalysis, adsorption, and advanced materials research. VNU-HUS ranked first in international publications in Vietnam in 2015 and organized the 46th International Chemistry Olympiad with 77 participating countries. VNU-HUS has also hosted a series of DAAD Summer Schools and international conferences on topics such as catalysis, sustainable development, and environmental planning.

Idea Behind RoHan

The creation of the RoHan Graduate School was a pioneering bottom-up initiative aimed at improving the current status quo and to interlink all partners over a long term. The idea behind RoHan was launched based on individual collaborations between Rostock research groups funded by several small grants. They were all motivated by the desire to share their expertise and experience with universities in Vietnam. The network enabled cooperation projects that could never have been undertaken before. Long-term projects involving several working groups could be realized, significantly advancing both individuals and the groups.

What Is Catalysis and Why Is It Important?

To understand why catalysis is an important and fast-growing area of research and how it relates to the Sustainable Development Goals, it is helpful to first explain some key terminology.

A chemical reaction is the process by which one substance (A) is transformed into another (B). For any chemical reaction to take place, a certain amount of energy—called "activation energy"—is required. This energy can be supplied by heat, electricity or light. In some cases, however, the amount of energy required can be considerable, or the reaction may not take place at all under the given conditions. A catalyst is a substance that reduces the activation energy required for a reaction to take place, often by providing an alternative reaction pathway. The result is an accelerated reaction, or even the ability for the reaction to occur when it wouldn't otherwise.

A useful metaphor for understanding catalysis is that of a traveller walking from point A to point B over a mountain. The mountain represents the activation energy and points A and B represent the starting material and the reaction product respectively. The path taken by the traveller symbolises the reaction pathway. If the traveller is alone, he can take the most direct but steepest route over the mountain (representing a catalyst-free reaction). However, if the hiker is accompanied by a local guide (the catalyst), the guide will help the hiker to take a less obvious, smoother path along the mountainside, which will require far less effort to reach point B.

Catalysis plays a crucial role in sustainable chemistry by making chemical processes more resource efficient, enabling reactions that are environmentally friendly or solve important social and environmental problems. In industry, more than 80% of chemical reactions involve catalysis. In the petrochemical sector alone, catalysis is essential in the refining of oil and gas, with 95% of all products involving catalytic processes. In addition, catalysis research is vital in many areas, including Processing and securing raw materials from both fossil and renewable sources; Developing sustainable energy supply chains, clean fuels and fuel cells; Synthesising new active ingredients for pharmaceuticals and agriculture to improve healthcare and increase crop yields for a growing world population; Addressing environmental concerns by developing methods for cleaning wastewater and treating industrial gases from combustion engines.

Given its enormous potential, expertise in both heterogeneous and homogeneous catalysis is critical to the well-being of humanity and the planet. Catalysis is central to several sustainable development programmes, particularly in developing countries such as Vietnam. Germany has been at the forefront of catalysis research and development for more than a century. As a result, there is a social responsibility to share these technologies with developing countries through academic exchange initiatives.

Fig. 1 Selected Sustainable Development Goals targeted within RoHan

Sustainable Development Goals Addressed by RoHan

Achieving the SDGs is an ongoing endeavour, initiated by the need to address specific global challenges. As mentioned above, catalysis plays a central role in driving chemical progress towards greater sustainability. Improving catalytic efficiency is key to making industrial processes more sustainable, environmentally friendly and cost-effective. As a result, SDGs that align with the teaching and application of catalysis science and have the potential to directly impact Vietnamese society have been identified as priorities (Fig. 1).

- **Goal 6: Ensure access to clean water and sanitation for all**. Clean, accessible water for all is an essential part of the world we want to live in, and there is sufficient fresh water on the planet to achieve this goal. Especially in Vietnam, waste and drinking water purification are among the major challenges. Therefore, catalysis topics related to new techniques for water purification, such as photocatalysis, are of utmost interest.
- **Goal 7: Ensure access to affordable, reliable, sustainable, and modern energy**. Energy plays a central role in nearly every major challenge and opportunity faced by the world today. Focusing on universal access to energy, increased energy efficiency, and the expanded use of renewable energy are crucial for creating more sustainable and inclusive communities and enhancing resilience to environmental issues like climate change. Technologies in catalysis that produce renewable energy and utilize alternative energy sources such as sunlight are being developed.
- **Goal 9: Build resilient infrastructure, promote sustainable industrialization, and foster innovation**. Investment in infrastructure is crucial for achieving sustainable development and empowering communities in many countries. It has long been recognized that growth in productivity and incomes, as well as improvements in health and education outcomes, are facilitated by infrastructure investment. Two laboratories are being installed in Hanoi in collaboration with RoHan, providing excellent facilities for research and education.

- **Goal 13: Take urgent action to combat climate change and its impacts.** Climate change is now affecting every country on every continent. National economies are being disrupted, and lives are being affected, with costs being incurred by people, communities, and countries today and potentially increasing in the future. Within the Rohan program, solutions for the use of biomass, particularly biowaste in developing countries, are being investigated. This effort aims to facilitate the transition toward a low-carbon economy.
- **Goal 4: Quality education.** By the end of 2016, half of the world's illiterate population was residing in South Asia. These concerning statistics highlight the absence of adequate learning environments and infrastructure in this developing region. In the RoHan project, efforts are being made to address this gap by establishing a research center to support the teaching of practical skills in chemistry at undergraduate/Bachelor and Master levels. This initiative is complemented by the introduction of a Double Degree Master program between the University of Rostock and the Hanoi University of Science and Technology.
- **Goal 5: Gender equality.** Despite global progress in gender equality indicators, women continue to be underrepresented at all levels of academic, industrial, and political leadership. RoHan aims to contribute to bridging this gap by providing fair learning opportunities to female students and researchers. RoHan's female scholarship holders are equipped with scientific tools and technical skills, empowering them to assume decision-making positions in scientific or environmentally oriented companies, academic institutions, or governmental bodies.
- **Goal 8: Decent work and economic growth.** While Vietnam experiences one of the steepest economic growth rates in the ASEAN region, there remains a necessity to increase employment opportunities, particularly among young people. RoHan aims to achieve this by training qualified researchers in both theory and practice. The program intends to provide young students and researchers with the skills necessary to access decent work and ultimately contribute to driving Vietnam's technological and economic growth.
- **Goal 17: Partnerships.** RoHan focuses on the implementation of a long-term strategic partnership with increased internationalization on both sides. This goal is being pursued through the establishment of the German-Vietnamese Catalysis Center GeViCat, which brings together stakeholders from academia and industry in Germany, Vietnam, and the ASEAN region.

2 Implementation of the First Phase: 2016–2020

The first phase of the RoHan project is implemented with a holistic view from the Master to the professorial level. As mentioned above, before 2016 only individual projects existed between the working groups at participating institutes. As part of the project's establishment, an intensive prior investigation trip to Vietnam was carried out, specially tailored to understand the relevant interdisciplinary project partners.

Topics particularly relevant to Vietnam and the stakeholders and cooperation partners were identified.

The status quo at the universities was analyzed, and relevant equipment selected. This was an important factor at the beginning of the project, as it represented the actual basis for exchange or joint learning. Based on this information, the exchange was built on four basic pillars (Fig. 2).

Bilateral Exchange Program

The first pillar is a comprehensive "Bilateral Exchange Program". Flexible short exchange stays of 1, 3, and 6 months were included for the training of Vietnamese Master and PhD students. This allowed for flexible and customized planning for the students. Subsequently, the Master theses were written and defended in English. Excellent Master students had the opportunity to complete their doctoral thesis in a bilateral mode (2.5 years in Vietnam, 6 months in Germany) or during a 3-year period entirely in Germany. Additionally, young Vietnamese scientists had the opportunity to conduct their own research during a one-year postdoctoral stay in Germany. Moreover, German scientists could support their studies with a 1- or 3-month stay in Vietnam.

During their stay in Rostock, the Vietnamese students were able to take part in the Graduate Academy of the University of Rostock free of charge. The Graduate Academy successfully supported young doctoral students, postdocs, and scientists at the University of Rostock in improving and developing their scientific and soft skills. The qualification program included, among other things, additional financial support for qualification measures. The Vietnamese students could also take part in language courses.

In addition, a South-South exchange between Asian universities at Master, PhD, and postdoc levels was initiated. The background to this was the establishment of the

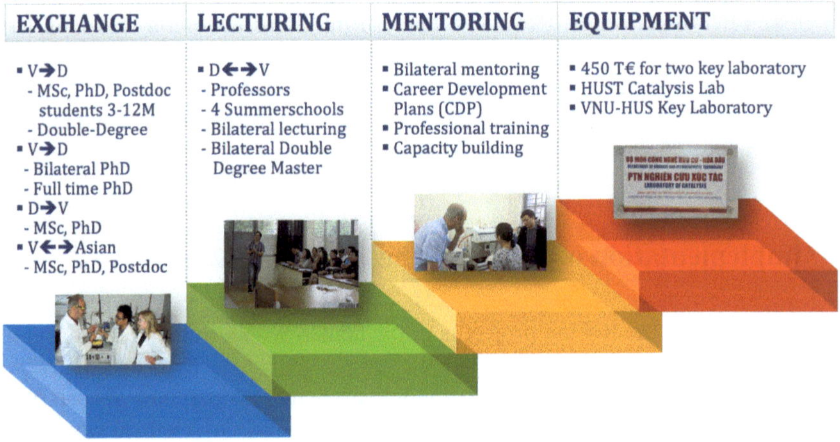

Fig. 2 The pillars of the RoHan project (V = Vietnam, D = Germany)

Graduate School at HUST and VNU-HUS in Vietnam. Furthermore, the network in Asia was strengthened, and the graduate school was made visible.

Bilateral Lecture Program

The second pillar is the lecture training of qualified Vietnamese and German academics via a "Bilateral Lecture Program." Annual winter and summer schools were held. The lecture program was launched as part of the summer school in Rostock in 2017. To ensure intensive networking and information exchange on the supported projects and among the stakeholders, summer schools combined with subsequent discussion and business meetings were held in Rostock (2017/2020) and Hanoi (2019). Extensive teaching, training, and project results were presented by the students at the graduate school and the professors. Furthermore, participants from politics, business, and civil society were invited to the discussion and business meetings.

Lectures were offered to HUST and VNU-HUS students. The offer was primarily aimed at Master and PhD students, as well as scientists at the postdoc level. As the number of post-graduate students at HUST and VNU-HUS was limited, selected teaching modules of the Graduate School were also offered to undergraduate students. This motivated Bachelor students to write their Master and PhD theses on the application of catalysis in chemistry, agriculture, and environmental sciences. Furthermore, targeted bilateral courses were held in Hanoi, such as "Introduction to Catalysis and the Environment," "Fundamentals of Environmental Science," and "Sustainable Chemistry." At Master and PhD levels, various lectures in the "Catalysis" modules were held as "lecture series." German and Vietnamese students, as well as professors, were trained together in information, education, and communication technologies (ICT) in Rostock before the COVID pandemic. The concept of e-learning in media and didactics was realized via "Inverted Classroom" courses. The teaching program was evaluated annually by students and lecturers. On this basis, the study program was revised and adapted to the Vietnamese Bachelor and Master program.

Bilateral Mentoring Program

During these exchange stays, special attention was given to mentoring the students through a "Bilateral Mentoring Program." Each research topic had one coordinator in Vietnam and one coordinator in Rostock. In cases of a bilateral/full PhD, each student had two supervisors (one from Vietnam, one from Germany). The research results were published in joint papers. For excellent students, especially women, individual Career Development Plans (CDPs) were developed and implemented, comprising Master, PhD, and/or Postdoc studies (e.g., Master in Hanoi, PhD in Rostock, Postdoc in Hanoi). Visits to companies in Germany and Vietnam were integrated into these plans. To strengthen the sustainability of the graduate school, an extensive program of skill-enhancing qualification courses was offered to Vietnamese and German PhD students, postdocs, and professors. Enhanced qualification courses included training in leadership (University of Rostock), entrepreneurship (Centre of Entrepreneurship Rostock), didactics (University of Rostock), and laboratory management/quality management/good laboratory practice (online, American

Chemical Society). Scientific exchange between active and former Bachelor, Master, and PhD students, postdocs, and professors was facilitated by a joint website for all supported projects. There, students found the corresponding contacts and partners. Graduates of the Bachelor, Master, and PhD programs were monitored regarding their further professional development. Excellent students were provided with various study opportunities and funding. Employment statistics of former graduates were collected, including an alumni network.

Equipment

A key focus of the first phase of the project was the installation of scientific equipment essential for conducting research in Hanoi. Resources such as chemicals and materials were also secured, significantly increasing research capacity. Research efforts in Vietnam were further supported by the provision of general laboratory equipment and small analytical instruments. To strengthen expertise, relevant staff and students were trained in Rostock. More than €450,000 was invested in scientific equipment during this phase, including a remarkable €170,000 for an EPR spectrometer. This instrument, which is unique in Vietnam, was an important foundation for future collaboration. The equipment facilitated the establishment of two dedicated synthesis and characterisation laboratories, laying the groundwork for advanced research activities (Figs. 3, 4, and 5).

Detailed Execution

The main objective of Phase I of the RoHan project was to establish a strategic partnership focused on education, capacity building in the field of catalysis and internationalisation of the partner universities. Quality education and the promotion of lifelong learning opportunities for both students and professors were central to this initiative. This phase laid the foundations for successful collaboration between research groups and created the conditions for effective use of resources to develop sustainable solutions in catalysis and resource management.

The project was initiated in May 2016 and formally submitted in June 2016. Following its approval in August 2016, all contracts between the University of

Fig. 3 RoHan Catalysis Lab at the HUST and new fumehoods for the synthesis of catalysts

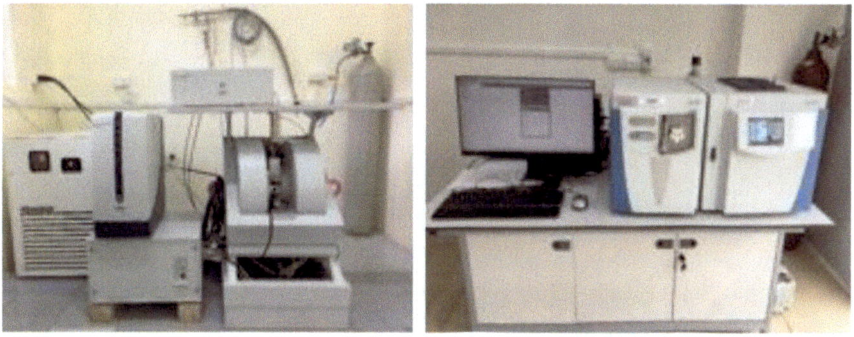

Fig. 4 New Catalysis Lab at the VNU-HUS and a photocatalysis reactor

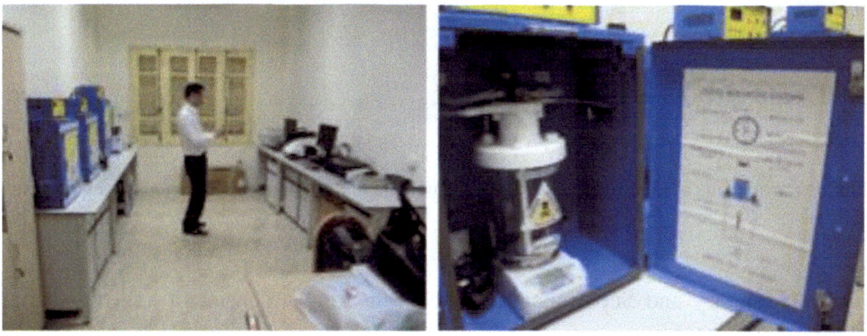

Fig. 5 Large equipment such as the EPR spectrometer as well as TPX equipment

Rostock (as the lead university) and the DAAD were finalised and signed. The total budget for the project was €2.2 million (€450,000 per year). As the funding included financial support for the whole of 2016, the initial activities of the project focused on its launch and the procurement of modern equipment for chemical laboratories in Vietnam. In November 2016, the first student selection process was completed and the first group of German and Vietnamese students joined their respective research groups, marking an important milestone in the promotion of international academic collaboration.

Two opening ceremonies, accompanied by interdisciplinary workshops, were held in Hanoi and Rostock (Fig. 6). These events provided an opportunity for the first working groups to meet, collaborate and engage in initial discussions on research directions. These meetings were instrumental in increasing the acceptance and visibility of the RoHan project. The visibility of the project was further enhanced by the creation of a dedicated website, as social media platforms were not widely used at the time. In the months that followed, the student selection process was refined, and initial criteria were established. Regular calls for applications were introduced, taking

Fig. 6 Participants of the RoHan Opening Ceremony at the Vietnamese—German Center in Hanoi (2016)

place in January and September each year, ensuring a consistent and transparent recruitment process.

The scholarship programme was managed flexibly, particularly in terms of visa issuance, which varied widely—from several weeks to just a few days. At this stage, the main criteria for awarding scholarships included a focus on research aligned with the Sustainable Development Goals (SDGs), the quality of the proposed research, and the academic qualifications of the students. A ranking system was introduced to ensure fairness and transparency. In parallel, significant investments have been made in laboratory infrastructure. Both major and minor equipment was purchased to quickly set up and prepare two laboratories for training. The equipment was state of the art and included a specialised device for measuring pore size, which is crucial for characterising catalysts and assessing their activity. Due to the specialised nature of this equipment, a Vietnamese student was trained to use it in Rostock before continuing his work with the equipment in Vietnam, ensuring its long-term and effective use. In the following years, the laboratory resources were further expanded, including the purchase of a €170,000 machine in 2019. Specialised training was provided in Germany and on site in Vietnam, with workshops to ensure ongoing skills development.

In May 2017, three Memoranda of Understanding (MoUs) were signed between all participating universities in Rostock and Hanoi, establishing formal links between the institutions. In 2018, an additional cooperation agreement was signed between the two Vietnamese universities. This was a notable achievement, as state dependency has historically limited opportunities for cooperation between Vietnamese institutions. To support the establishment of long-term, strategic, university-wide partnerships, several cooperation meetings were held between the administrations of the University of Rostock and HUST. From the outset, the depth of cooperation exceeded that of typical projects. This bottom-up approach to research and cooperation facilitated a parallel top-down process within the universities, enabling the acquisition of additional projects beyond catalysis research. The integration of intensive research and meaningful exchanges attracted attention far beyond the region, as evidenced by visits and interest from high-level politicians e.g. from Japan, Singapore, Brunei. German colleagues initially approached the collaboration from their own perspectives and experiences, which were understandably limited. In order to gain insight into the specific challenges in Vietnam and to tailor applied research to local needs, a "Future Solution Workshop" was organised in Rostock. This workshop brought together around 25 German and Vietnamese students and professors to analyse key areas where catalyst research could have an impact. Discussions highlighted critical issues, including the high rates of disease caused by inadequate healthcare infrastructure. Contaminated drinking water, hospital-acquired infections and severe air pollution were identified as major contributors to these challenges. These insights proved invaluable in shaping future collaborations.

In the actual Vietnamese context, the catalytic "needs" to be fulfilled spread through several industrial sectors and attain various environmental and energetic issues including:

- **Catalyst for water treatment, gas purification and post-process gaseous waste treatment** e.g. Catalyst supports and adsorption materials, i.e. nanomaterials, sieve-like materials, mesoporous materials, metal oxides as well as (photo)catalysts doped with transition metals and main-group elements.
- **Catalysis for chemical and energy conversion** e.g. selective oxidation of alkanes, oxidative dehydrogenation of n-butane, Fisher-Tropsch synthesis of hydrocarbons, catalytic cracking and esterification for the production of biodiesel. These technologies aim to cover the increasing demand of hydrocarbon fuels for mobile applications, parallel to the development of hydrogen and battery-based alternatives.
- **Catalysts for mitigation of the effects of human activities on the environment** e.g. sequestration of heavy metals on soil and water, complete oxidation of aromatic hydrocarbons, hydro dechlorination of tetrachlorethylene, oxidation of CO and reduction of NOx in the atmosphere and sustainable organic synthesis.
- **Sustainable Materials** e.g. use of catalytic approaches for the modification of bio-based polymers and for the synthesis of novel bio-derived and/or bio-degradable polymeric materials.

In line with the above topics, annual summer schools were held in 2017, 2018 and 2019 in both Germany and Vietnam. Due to the pandemic, these were replaced by online winter schools in 2020. While the pandemic significantly disrupted exchange activities, it also opened up new opportunities for digital collaboration and university partnerships (see Sect. 8).

During the summer and winter schools, participants not only attended lectures, but also engaged in non-academic networking with small and medium-sized enterprises (SMEs) and large corporations. This broadened students' perspectives and provided valuable industry experience. For the professors, collaboration with non-university partners led to the initiation of several large-scale projects, particularly in the wind power and energy sectors, many of which were based on the RoHan project. In addition, some students were able to secure employment after completing their studies, further expanding the impact of the project. Networking within the RoHan alumni community was a priority from the outset, with various excursions and meetings organised each year to encourage continued engagement. Scientific seminars were held online to present and share current research. These activities played a crucial role in helping participants feel a strong sense of identification with the project and its overarching goals. This connection was essential to ensure a lasting impact on Vietnamese society.

Close collaboration between funders and universities is essential for the successful implementation of projects of this scale. Annual collaborative meetings have been essential in addressing and resolving challenges related to equipment funding, as well as fostering a common understanding of the projects, their participants and their added value. Establishing links with other Research Training Groups was particularly beneficial as it facilitated interdisciplinary exchanges and learning processes. This proactive approach allowed potential challenges to be identified and addressed at an early stage, ensuring smoother project implementation.

3 Outcomes of RoHan Part 1

As outlined above, the RoHan project has successfully transformed the previously unstructured collaboration between the Vietnamese partners into a systematic, strengthened and well-networked partnership. At the beginning of the project, comprehensive Memoranda of Understanding (MoUs) were signed by all partners, laying the foundations for effective collaboration. These agreements facilitated the exchange of teaching staff, the recognition of academic achievements and the joint organisation of courses. The advertising and acceptance of the students as well as the increasing enthusiasm of the students and professors led to an extraordinary exchange. Expectations were far exceeded. By the end of 2019, 66 Vietnamese and German students and researchers had received scholarships through the programme. In particular, the project also supported 13 students from ASEAN countries, including India, Malaysia and Indonesia, to conduct catalysis research in Hanoi.

Most of the students were re-employed at the universities before and after the exchange. The establishment of two partly equipped research laboratories at the Vietnamese partner universities played a very important role here. State-of-the-art scientific equipment was the key to continuing research at the same level. In addition, eight PhD students (five in Germany and three in Hanoi) successfully completed their studies by the end of the funding period, underlining the substantial academic contributions of the project. The productivity of these exchanges and research activities is evidenced by the publication of over 63 papers in peer-reviewed scientific journals, and 34 posters and oral presentations at international conferences and symposia. These contributions resulted from the research efforts of Vietnamese scholars in Rostock and Hanoi, as well as South-South exchange students studying in Hanoi.

Through the annual summer schools, workshops, lectures, training courses and seminars, many students have been educated and made aware of the RoHan project. In particular, the importance of catalysis technology was made clear to the students. The number of Masters and PhD students increased dramatically. An important milestone was the establishment of a Double Degree Master's programme in Sustainable Chemistry between the University of Rostock and HUST, which further cemented the partnership. Due to the different degree programmes at the universities, the degree programmes could only be introduced and mutually recognised at the end of the first funding period after a long period of negotiation. The most difficult part was the low level of acceptance on the German side at the beginning of the project. The acceptance increased as more and more students came to Germany and as the professors became more aware of the potential of the exchange. The time-limiting factor was the administrative hurdles in Germany. The Vietnamese side was very flexible in this respect, as the courses were reorganised and adapted during this period. There was a great demand for new topics to be included in the curriculum. Once the contracts were signed, the exchanges took place. To describe this in more detail, students are exchanged of one year during their Master studies. The first two semesters are spent at the home university. After that, the students attend lectures and write their Master's thesis at the university in the guest university.

Another important outcome was the structural impact on the partner institutions in Hanoi, driven by the increased autonomy of Vietnamese universities. New governance models, such as a university council, have been implemented via the German especially Rostock Model. Here an extensive exchange of ideas through the integration of academic and administrative staff was achieved. In addition, the internationalisation strategies of the Vietnamese universities and the University of Rostock were significantly strengthened, resulting in 17 mono-, bi- and multilateral national and international projects with a total funding volume of over €5 million.

In summary, the first phase of the RoHan project has served as a beacon, catalyst and driving force for the establishment of a robust and productive partnership between Germany and Vietnam, closely aligned with the objectives of SDG Goal 17.

However, only a few truly long-term personal relationships were established. Although some Vietnamese students had joined the Vietnamese universities, it was not possible for them to build up their own research. These people were often employed as lecturers. Likewise, the double degree was only introduced with a few people at the end of the year. To improve this, new strategies were introduced, which are explained in detail in the second phase.

4 Realisation of the Second Phase: 2021–2025

In the second phase of the RoHan Rostock-Hanoi DAAD SDG Graduate School, the focus shifted to ensuring sustainability. A key initiative was the establishment of the German-Vietnamese Catalysis Centre (GeViCat) based at HUST. This new institution was designed to have a lasting and sustainable impact in Hanoi and the wider ASEAN region, with the aim of enabling the host institution to achieve full independence as soon as possible (see Sect. 7.1).

A key element of this phase was the introduction of pre-research initiatives, new starting grants and the double-degree master's programme. Pre-Research explored and evaluated innovative ideas at an early stage, providing a solid foundation for future projects. The new starting grants provided financial support to promising postdoctoral researchers, helping them to advance their research projects and careers. The Double Degree Master's Programme allowed students to study in both Rostock and Hanoi and to obtain two degrees upon graduation. This programme fostered intercultural cooperation and knowledge exchange between the two universities, further strengthening academic ties. In addition to promoting scientific exchange and education, the second phase enriched existing structures by developing an active alumni network and establishing an interdisciplinary SDG Graduate School Alliance. These initiatives have further strengthened collaboration and ensured that the impact of the project will extend well beyond its immediate duration.

German-Vietnamese Catalysis Research Centre - GeViCat

Building on the achievements of the first phase, additional measures were implemented to enhance research capabilities and strengthen cooperation between Rostock and Hanoi. Efforts were also made to optimise the overall research environment. This included the development of modern, well-equipped workspaces where researchers could carry out their projects. In particular, the 400 m^2 GeViCat laboratory was established to consolidate all research equipment in one central location. This significantly increased research capacity by facilitating access to state-of-the-art equipment and advanced technologies (Fig. 7).

At GeViCat, research focuses on topics of highest relevance to the Vietnamese and Southeast Asian context, with an emphasis on developing, applying and addressing local challenges. GeViCat has become a leading innovation hub in Vietnam and a beacon for scientific cooperation and development across Southeast Asia, providing a robust scientific infrastructure. The centre has created a collaborative environment that fosters partnerships between universities, research institutes and companies across the region. The establishment of GeViCat has strategically consolidated previously scattered research activities and created a centralised platform to address pressing local issues. A key focus has been to promote balance between the partners by leveraging their complementary expertise and resources. This approach fosters a true university partnership based on mutual benefit and shared goals. With its state-of-the-art laboratories equipped with modern scientific tools, GeViCat has become a premier research facility for students and researchers not only in Vietnam but in

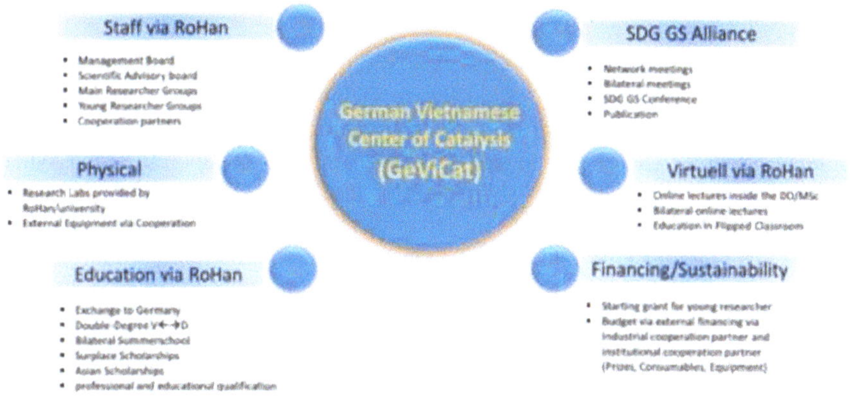

Fig. 7 Structure of the newly formed German Vietnamese Center of Catalysis

the wider Asian region. It has also expanded its network of academic and industrial collaborators, enabling research groups from different institutions to share analytical equipment and expertise. This collaborative framework has created a 'win–win' situation for all partners involved, maximising resources and outcomes. GeViCat's networking initiatives are further supported through participation in conferences, workshops and joint calls for research proposals. In addition, the Centre actively facilitates South-South student exchanges, reinforcing its role as a hub for regional cooperation and knowledge sharing.

GeViCat facilitated intensive collaboration with regional partners in Germany, including the Interdisciplinary Faculty of the University of Rostock, the Leibniz Institute for Catalysis (LIKAT), and small and medium-sized enterprises (SMEs). Similarly, strong partnerships have been established in Vietnam, such as with the Vietnam Petroleum Institute and the Dung Quat Refinery, to promote the development of joint projects. There has also been a focus on expanding and strengthening links with new academic partners in Vietnam and across Asia. The state-of-the-art equipment and laboratory facilities at GeViCat served as a strong incentive to initiate collaborative research and cooperative projects. These efforts were supported not only by the active engagement of scientists, but also by the involvement of key stakeholders, including NGOs, SMEs and government agencies. This multi-stakeholder approach ensured that GeViCat became a central hub for innovation and collaboration in the region.

Building on the foundation of the RoHan DAAD SDG Graduate School, GeViCat became a cornerstone of the sustainability strategy for the strategic university partnership between the University of Rostock and its Vietnamese partners. The long-term goal is to establish a robust interdisciplinary network of collaborative projects in Vietnam, while fostering a truly global and equitable partnership between the University of Rostock, LIKAT, HUST and VNU-HUS.

SDG Graduate School Alliance

Building on the success of previous collaborative sessions held in Bonn, Germany (2017), Hawassa, Ethiopia (2018) and Hanoi, Vietnam (2019), as well as joint activities such as the development of research proposals during Phase I, the seven SDG Graduate Schools (SDG GS) agreed to strengthen their collaboration and synergies in Phase II through the establishment of the 'SDG GS Alliance'. The SDG GS Alliance was led by a transversal group of researchers representing all participating SDG Graduate Schools. The Alliance served as an important platform for dissemination and networking, significantly strengthening the education sector of all partnering higher education institutions (HEIs) and contributing to the achievement of the UN 2030 Agenda. It provided extensive opportunities for global collaboration, connecting academics at all career stages with external stakeholders. The Alliance facilitated the joint development of solutions to SDG-related challenges, through inter- and transdisciplinary research and collaborative initiatives. Firmly rooted in the belief that the higher education sector has a critical role to play in advancing sustainable development globally, the SDG GS Alliance consolidated and systematised its collective expertise in the design and management of international development projects. By identifying and sharing best practices in higher education partnerships for sustainable development, the Alliance has established a model for impactful global collaboration.

5 Integration of German and Foreign Scientists and Academics

As a leading graduate school promoting cooperation between Germany and a rapidly developing country like Vietnam, RoHan Phase II recognised its significant responsibility to society, especially to its students. In order to maintain the highest standards of education and knowledge transfer, Phase II focused on capacity building at both partner universities. A cornerstone of these efforts was the introduction of the bilateral Double Degree Masters Programme, a key achievement of Phase I. In addition, online and virtual lectures were introduced, as well as advanced practical courses using state-of-the-art equipment acquired through RoHan. These lectures were mainly delivered by GeViCat faculty and alumni, ensuring the integration of expertise and continuity of training.

In order to promote networking and information exchange between the supported projects and stakeholders, the double-degree Masters and PhD programmes were complemented by five annual RoHan Summer Schools, held alternately in Hanoi (2022, 2024) and Rostock (2021, 2023, 2025), as well as an SDG DAAD Graduate Conference (2023). These activities have been further enriched by participation in international conferences, broadening the exposure and connections of students and faculty. To ensure the sustainability of the graduate school, a comprehensive capacity building programme was developed for Vietnamese and German PhD students, postdocs and professors. This programme included training in leadership (University of Rostock), entrepreneurship (Centre of Entrepreneurship Rostock), didactics (University of Rostock) and training in laboratory management, quality management and good laboratory practice (GLP) (provided by Klinkner & Partner). Vietnamese students studying in Rostock were also given free access to the University of Rostock's Graduate Academy, which provides comprehensive support to young PhD students, postdocs and researchers to improve their scientific and soft skills. This qualification programme included financial support for qualification measures and offered additional opportunities such as participation in language courses.

During these exchanges, particular emphasis was placed on the provision of robust mentoring for participating students, guided by the 'leave no one behind' strategy. Each exchange was supported by a structured mentoring programme to ensure personalised guidance and support. For each research topic or collaboration, a dedicated coordinator/contact person was appointed in both Vietnam and Rostock, as established in RoHan Phase I. In the case of bilateral or full PhD programmes, each student had two supervisors—one from Vietnam and one from Germany. To maintain progress and accountability, students submitted small scientific reports to the GeViCat Coordination Board and their supervisors every six months. These research efforts often culminated in joint publications showcasing the collaborative outcomes of the High-performing Masters students were also given the opportunity to complete a full PhD in Germany and Vietnam. After the PhD, these scholars were supported with a small 'start-up grant' for six months to help them launch their independent careers. During this period, they were encouraged to develop their own projects and secure funding e.g. via local research funding such as Ministry of Science and Technology (MOST), Vietnam Academy of Science and Technology (VAST), National Foundation for Science and Technology Development (NAFOSTED) or via Vingroup Innovation Foundation (VinIF). Vietnamese PhD students were also offered six months of preparatory training in Vietnam before starting their research stays in Germany. During this phase, efforts were also made to deepen cooperation between Vietnam's two leading universities and smaller regional institutions, fostering links through initiatives such as the South-South Exchange Programme.

The exchange of scientific knowledge between current and former Masters and PhD students, postdocs and professors was facilitated by the website, which presented all funded projects. This platform provided an accessible directory of contacts and collaboration opportunities.

In addition to research mentorship, students were offered additional funding to support their studies before and after their research stays, as well as grants for young researchers. These measures were in line with the inclusion policies of both partner universities in Hanoi, ensuring accessibility for people with financial difficulties or disabilities. For example, all RoHan laboratories were designed to be wheelchair accessible. In addition, students participating in the Double Degree Masters Programme benefited from tuition fee waivers at their home universities during their exchange period.

6 ICT and Digitalisation Inside RoHan—Lessons Learned

The academic world, particularly in experimental fields such as chemistry, has undergone profound and transformative changes as a result of the Covid-19 pandemic. Historically, these disciplines have emphasised face-to-face interaction and practical experience as essential components of education and research. However, the landscape of academic exchange and learning has evolved significantly over the past five years. Prior to 2019, the RoHan Graduate School was fundamentally structured around physical presence. The curriculum relied heavily on face-to-face lectures, with no provision for online learning. Practical courses, which were central to providing hands-on experience, further reinforced the need for physical presence. The exchange of knowledge and skills between Masters and PhD students was mainly facilitated through summer schools and specialised courses. These were typically delivered as intensive block seminars scheduled after the main lecture periods. While effective in fostering collaboration and in-depth learning, this traditional format inherently favoured individuals and institutions with greater financial resources. Participation in such educational opportunities often required substantial funding for travel, accommodation and other associated costs, creating barriers for those with limited budgets.

Communication and Teaching

The RoHan Graduate School's transition to digitisation in communication and education marked a significant turning point, particularly in its approach to teaching. This rapid change was exemplified by the adoption of online teaching, moving away from traditional face-to-face methods. Key milestones included the launch of the *Introduction to Sustainable Chemistry* course and the continuation of bilateral exchange programmes via digital platforms, demonstrating the school's adaptability. By removing the need for physical attendance, the transition expanded opportunities for participation and made education more accessible. To support this shift, standardised courses were developed to enable the seamless transfer of knowledge from professors to students. Digital tools facilitated the exchange of data and lecture materials, enabling professors to teach independently. Platforms such as Zoom, Microsoft Teams, BigBlueButton and even smartphones played a key role and formed the backbone of these new teaching methods. One innovative teaching approach was the integration of hybrid learning, combining face-to-face lectures with online sessions, including the 'flipped classroom' model. This method combined high quality research and theoretical instruction with hands-on practical training on equipment. The recording of lectures and the ability to revisit them on demand transformed the learning experience, allowing students to engage with the material at their own pace. Interactive elements such as assignments and discussions further increased student engagement and understanding.

The shift to digital education also increased the number of students participating in each module, enabling collaboration between several universities and fostering deeper partnerships. This expansion contributed significantly to the internationalisation of the student body at the University of Rostock. Digital platforms also allowed the school to connect with its extensive alumni network, extending its global reach. For example, the 2019 and 2020 winter schools were conducted online. While previous face-to-face sessions typically attracted 20–25 participants, mostly Vietnamese colleagues, the online format attracted over 100 participants from around the world. This inclusivity allowed the school to engage alumni regardless of their location. In addition, unused travel funds were reallocated to purchase additional equipment for the GeViCat Centre, thereby increasing its resources.

However, the transition was not without its challenges. A major hurdle was the initial resistance to online education, due to language barriers and scepticism about digital platforms. Overcoming these barriers required a strong emphasis on digital literacy—equipping staff, students and mentors with the skills to effectively use technology for learning, teaching and communication. These skills ranged from technical skills and critical thinking to reading and writing in digital environments. Interestingly, the Vietnamese participants showed greater flexibility in adapting to these changes than some of the German teachers. To address these challenges, the school implemented digital training programmes for all stakeholders to ensure they could navigate the new digital landscape with confidence. This comprehensive approach not only addressed the initial difficulties, but also set a new benchmark for academic excellence, inclusivity and innovation.

Data and Information

Traditionally, data and information management has relied on paper laboratory notebooks and individual storage of research data. However, managing the entire lifecycle of research data has become a critical aspect of school operations. This lifecycle includes data generation, processing, enrichment, archiving and publication, ensuring the integrity, accessibility, and utility of the data at every stage. These efforts reflect the school's commitment to innovation and adherence to best practice. With the implementation of Good Laboratory Practice (GLP) standards, the RoHan Graduate School has taken significant steps to modernise its data and information management strategies. By integrating the principles of FAIR (Findable, Accessible, Interoperable, Reusable) and CARE (Collective Benefit, Authority to Control, Responsibility, Ethics), the school ensures the efficient, ethical and socially responsible management of research data. This approach not only maintains scientific rigour, but also promotes transparency and inclusivity in data management practices.

Some research groups at the University of Rostock have been at the forefront of introducing and testing e-journals such as ELabFTW and ChemMotion, and extending their use to developing countries. This innovative initiative has paved the way for modernised data management practices in diverse settings. However, the process has not been without its challenges. The lack of uniformity between different e-lab systems has required ongoing digital training for staff, mentors and students to ensure seamless implementation and effective use. Complementing these advances, the integration of online safety lectures from the American Chemical Society (ACS) highlights the growing importance of online teaching and the need for robust digital education in the sciences. The RoHan Graduate School has further enriched the learning experience by organising monthly interactive activities. These sessions, focusing on topics such as data management, teaching methods, and poster presentations, have fostered active engagement and collaboration among participants.

Critical Reflections

While digitalisation has greatly increased the efficiency and reach of communication, it has also introduced critical challenges, particularly in the area of communication skills. Effective communication is inherently multi-dimensional, relying on elements such as gestures, tone of voice and non-verbal cues—many of which are diminished or lost in digital formats. This reduction in interpersonal nuance, combined with limited opportunities for international exposure and cultural immersion, can hinder the development of intercultural competence. For both students and teachers, this shift can lead to reluctance or reduced confidence in networking, ultimately weakening social skills and the ability to build social capital. As a result, the formation of meaningful professional relationships and opportunities for collaboration may suffer. Furthermore, digital environments often obscure the underlying human elements, such as shared goals and personal loyalty, that are essential for fostering trust and collaboration in academic and professional settings.

Another challenge is the significant time and financial resources required to prepare and deliver effective digital communications. From creating engaging content to ensuring clarity and accessibility, the demands of digital platforms can put significant pressure on already limited resources. Addressing these challenges is essential to ensure that the benefits of digitalisation do not come at the expense of critical interpersonal and cultural skills.

The importance of sound data management in the digital age cannot be overstated. Often referred to as the 'currency of the future', data brings with it a host of ethical and practical dilemmas. The increasing visibility and potential mutability of data, driven by advancing technologies, raises critical concerns about data integrity, interpretation and misuse. Questions of data ownership and accountability are becoming more pressing, particularly as hosting platforms may claim rights over the data they store or share. Furthermore, the lack of well-defined ethical guidelines for data management and use highlights the evolving and unresolved nature of these challenges. These issues call for a forward-looking approach to data governance that prioritises ethical principles, transparency and responsible use to guard against misuse.

In academia, the transition to digitisation has not been universally embraced, particularly by traditionalists within the scientific community. This resistance underscores the need for experienced professionals who can bridge the gap between conventional and digital methodologies. Digital formats, while efficient, often reduce opportunities for face-to-face interaction and discussion—key components of effective teaching and learning. The lack of established ethical guidelines in digital education exacerbates these challenges, creating an urgent need for frameworks that ensure equity and integrity in digital learning environments.

In addition, digital resources and opportunities are not equally accessible to all students, exacerbating inequalities in learning and creating barriers for those without sufficient access to technology. A related concern is the potential erosion of basic skills such as critical reading, articulate speaking and effective information retrieval. Over-reliance on digital tools risks fostering complacency and weakening the very skills that underpin academic and career success. Addressing these challenges requires a balanced approach that embraces the benefits of digitalisation while preserving essential human skills and promoting inclusiveness in access and opportunity.

Impact on the Sustainability

The reflections in the previous section underscore the importance of a balanced approach to digitalisation that not only acknowledges its complexities and challenges, but also harnesses its transformative potential. The digital transformation of the RoHan Graduate School represents a critical opportunity to promote sustainability in academia and beyond. By seamlessly integrating the strengths of online and face-to-face learning, the school can cultivate a more inclusive, innovative and sustainable educational environment. The hybrid model of education, combining online and face-to-face learning, transcends geographical and cultural boundaries and opens up new opportunities for collaboration and knowledge sharing. This approach promotes the development of transferable skills such as digital literacy, interdisciplinary problem solving and effective communication, equipping students with the tools they need to thrive in diverse professional environments. It also facilitates the efficient transfer of knowledge and expertise from academia to industry, supporting practical applications and fostering a culture of lifelong learning. Exploring innovative learning formats—such as 360° video demonstrations of experiments, virtual reality (VR) experiences and educational podcasts—has the potential to revolutionise the way students engage with complex concepts in chemistry and related disciplines. These immersive and interactive methods cater to different learning styles, enhance comprehension and improve knowledge retention, making education more dynamic and accessible. The RoHan Graduate School's commitment to diverse teaching methods is reflected in its initiatives such as personalised video introductions by alumni and the international expansion of E-Labs to locations such as Vietnam.

A notable benefit of digitalisation is its positive impact on the environment. By reducing the need for extensive air travel for academic purposes, the school is actively contributing to the reduction of climate emissions. The digital model also broadens participation, enabling more students and professionals from different locations to participate in educational programmes, thereby extending the school's global reach while minimising its carbon footprint.

7 Leave No One Behind Strategy Inside the RoHan Program

The 'leave no one behind' strategy is one of the core principles of the 2030 Agenda for Sustainable Development and the SDGs. From the beginning, this principle was actively implemented in the RoHan project. As mentioned above, special attention was given to student monitoring and career development planning. The aim was to ensure that all students—regardless of gender, geography, age, socio-economic background or availability of resources—had equal opportunities to succeed.

One of the key strengths of the programme was its inclusivity, offering opportunities to students who may not have met the higher English language requirements of many European or American Masters programmes. By supporting these students and giving them access to a high-quality education at a leading German university and research institute, RoHan demonstrated the value of this approach. The results speak for themselves: many of these students went on to successfully complete PhD or post-doctoral studies in Europe, while others secured competitive positions in Vietnam. In addition, RoHan created pathways for students from small towns in Vietnam and those facing financial or family difficulties, ensuring that they had the opportunity to further their education. Their studies were fully funded, allowing them to focus on their academic and professional development without financial burdens. Without RoHan's support, many of these students would not have had the opportunity to pursue Master's and doctoral degrees.

The integration of a gender perspective in chemistry, particularly within the RoHan project, was also a crucial component. Despite the traditionally low representation of women in the chemical sciences, promoting gender balance was a priority in the selection process. Encouraging women's participation in chemical research was not only a matter of equity, but also essential for promoting diverse perspectives and fostering scientific innovation. As a result, more than 40% of all RoHan students were women.

The selection process for the scholarships was entirely merit-based, regardless of gender or ethnicity. To ensure fairness and inclusiveness, on-site interviews and direct assessments were conducted with all applicants before the final selection was made. This approach allowed for a more comprehensive assessment of candidates, taking into account their academic achievements, research potential and personal skills. Through individual assessments, recommendations and guidance, RoHan ensured that each applicant, regardless of background, had an equal opportunity to participate and succeed in the programme.

The RoHan project actively encouraged interdisciplinary research, recognising that chemistry does not exist in isolation but thrives in collaboration with other scientific disciplines. This inclusive approach ensured that funding was allocated on the merits of research ideas rather than specific subject groups or individuals, allowing for diverse and innovative projects. In particular, the project also introduced novel training approaches for chemistry teachers, extending its impact beyond traditional research. Collaboration extended to the Interdisciplinary Faculty, including chemistry, biology and physics, reinforcing the idea that scientific progress benefits from cross-disciplinary exchange. This interdisciplinary framework facilitated the emergence of new research directions and pioneering methodologies, particularly with a focus on Vietnam, contributing to the advancement of chemistry as a field.

In the area of digitisation, ensuring wide online access to RoHan's resources and research outputs was a cornerstone of the project's success. More than 80% of the publications were made available as open access, enabling global participation without geographical or time constraints. This commitment to transparency and accessibility not only improved the dissemination of scientific knowledge, but also strengthened the global research community, fostering international collaboration and sustainable chemistry practices.

8 Summary, Impact and Legacy

The DAAD and BMZ-funded RoHan Graduate School was a pioneering initiative in both academia and industry. A key achievement of the project has been the strengthening of collaboration between the four participating institutions, fostering deeper cooperation and an extensive research network. This synergy has led to the acquisition of major follow-on projects with a combined funding volume of over €10 million, significantly enhancing the financial and research capacity of the universities involved.

A particularly notable outcome has been the development of the Vietnamese universities into leading research institutions in the field of chemistry. This progress is reflected in more than 80 published research papers, underlining their growing academic influence. In addition, the state-of-the-art equipment procured through the project has not only facilitated these publications, but also serves as a foundation for future research efforts. As part of the German-Vietnamese Centre of Catalysis, these resources will continue to expand, opening up new opportunities for further research and project acquisition. Another notable impact of the RoHan Research Training Group is its alumni network. Many former students have secured influential positions in academia, establishing their own research groups and driving innovation in the scientific community. Beyond academia, the long-term impact of the project is also evident in the industrial sector. Small and medium-sized enterprises (SMEs) from Germany have successfully entered the Vietnamese market, employing RoHan graduates and contributing to the transformation of the country's chemical industry. These developments are bringing Vietnam closer to its goal of becoming a global leader in sustainable chemical practices.

An important legacy of the project is the establishment of the German-Vietnamese Centre of Catalysis (GeViCat), which will serve as a central hub to promote collaboration between German and Vietnamese researchers and institutions in the field of catalysis. Upon completion of the project, GeViCat aims to foster long-term partnerships by integrating permanent research groups specialising in different aspects of catalysis. These groups will work closely together on innovative, forward-looking projects, using their combined expertise to develop new insights and breakthroughs. This collaboration will facilitate an intensive exchange of knowledge and maximise the efficient use of existing resources.

In addition to strengthening academic cooperation, the long-term vision of GeViCat is to establish a strategic partnership between the University of Rostock, the Leibniz Institute for Catalysis and the two Vietnamese universities—a partnership that goes beyond traditional cooperation. This initiative will not only benefit students but will also deepen institutional ties and expand the RoHan network to include entire universities in both Rostock and Hanoi.

But what might overarching and independent support for students and institutes look like? RoHan Coordinators started thinking about this very early on. One idea for ensuring sustainable research funding is the formation of a so-called association. This foundation can be a strategic step towards securing long-term funding and the sustainable success of scientific cooperation between Europe and Asia. Normally, research projects such as the Rohan project are mainly financed through temporary programmes such as DAAD funding. However, they offer a dependency on funding rather than a long-term alternative. This can be circumvented by an independent organisation. It can tap into alternative sources of funding to secure long-term projects. It has flexibility and independence because, unlike a university research institution, it can make unbureaucratic decisions and is not bound by state regulations.

The idea is to build a bridge between the university and industry, whereby these can be linked in a targeted manner. Financially, the association can be financed by membership fees or research funding. It is also easy to involve industry, which can be involved through sponsorship models.

The currently planned 'Europe Asian Chemistry Excellence Cluster Association' could play a central role in the international research landscape and thus take on the core tasks independently of the funding institute. These are the promotion of students and young scientists, the organisation of interdisciplinary seminars and workshops and access to modern research equipment. Members are given access to each other's state-of-the-art laboratories and equipment. This facilitates research as well as bilateral publications. This lays an essential foundation for long-term scientific co-operation. This is currently an idea of the authors and will be refined and realised over the next few years.

In summary, the RoHan Graduate School has had a profound and lasting impact on academic advancement, industrial development and sustainable research in Vietnam. Its legacy is firmly embedded in the establishment of GeViCat, the strengthening of international research networks, and the creation of long-term academic and industrial collaborations. By fostering innovation, knowledge exchange and interdisciplinary research, the project has not only empowered the next generation of scientists, but also laid a strong foundation for continued progress in science and industry. Through its alumni, institutional partnerships and ongoing research initiatives, RoHan's impact will continue to drive sustainable development and scientific excellence well into the future.

References

Bösch F, Su PH (2018) Invisible, successful, and divided: Vietnamese in Germany since the late 1970s—WIDER Working Paper 2018/15. https://doi.org/10.35188/UNU-WIDER/2018/457-5

Cordova KE, Yaghi OM (2019) Building a global culture of science—the Vietnam experience. Angew Chem Int Ed 58(6):1552–1560. https://doi.org/10.1002/anie.201812076

DFG-NAFOSTED Cooperation (2024) https://www.dfg.de/en/about-us/international-cooperation/news/reports/2024/nafosted-2024-2025

Nguyen TV, Ho-Le TP, Le UV (2017) International collaboration in scientific research in Vietnam: an analysis of patterns and impact. Scientometrics 110(2):1035–1051. https://doi.org/10.1007/s11192-016-2201-1

Dirk Hollmann studied Chemistry at the Technical University of Berlin (Prof. Siegfried Blechert). He completed his doctorate at the Leibniz Institute for Catalysis (LIKAT) under Prof. Matthias Beller. For Postdoc research Dirk joined Prof. Karl Anker Jørgensen at the Center for Catalysis, Aarhus University as well as Prof. Angelika Brückner at LIKAT. Since 2018, he has been leading the independent "Sustainable Chemistry" research group (Mentor: Prof. Udo Kragl) at the University of Rostock. Since 2024 he is included in the new developed Business Chemistry study program. His research interests include the development of environmentally friendly processes based on the principles of circular economy. Dirk and Esteban are the initiators of the RoHan SDG Graduate School.

Esteban Mejía is an experienced chemist and group leader at the Leibniz Institute for Catalysis (LIKAT) in Rostock, Germany. Originally from Colombia, he studied chemistry at the National University in Bogotá. In 2008, he moved to Switzerland to pursue doctoral studies at ETH Zurich under the guidance of Prof. Antonio Togni, specializing in homogeneous catalysis. Afterwards he joined the LIKAT in Rostock as a postdoctoral researcher in Prof. Matthias Beller's group. Since 2014, he has led his own research group at LIKAT, with a focus on catalytic processes in polymer chemistry. He obtained his habilitation in 2020. His research covers a wide range of topics, including organometallic chemistry, sustainable catalysis, silicon-based compounds, and photochemical transformations.

Tabea Angela Thiel is a German Scientist in Chemistry currently doing her Ph.D. at the Leibniz Institute for Catalysis (LIKAT) in Rostock, under the supervision of Prof. Dr. Esteban Mejía. She obtained her bachelor's degree at the Freie Universität and her master's degree at the Technische Universität Berlin. Her research focuses on catalytic polymerizations, reaction engineering, and heterogeneous (photo)catalysis. She is expected to complete her doctoral thesis by 2025.

Udo Kragl is a German chemist and biotechnologist specializing in biocatalysis and green chemistry. He studied chemistry at the University of Bonn, where he earned his Ph.D. in 1992 and later completed his habilitation in Technical Chemistry. Since 1998, he has held a professorship at the University of Rostock. His research focuses on enzyme-based synthesis, membrane technology, ionic liquids, and process development for chemo- and biocatalytic systems. He has published over 150 scientific papers, holds 20 patents, and has supervised more than 40 Ph.D. students. He is the founding Dean of the university's Interdisciplinary Faculty.

Le Minh Thang is a leading Vietnamese scientist at Hanoi University of Science and Technology (HUST), where she heads the Applied Catalysis research group in the School of Chemistry and Life Sciences. Her work explores a range of topics in catalysis, including the oxidation of hydrocarbons like propane and propylene into acrolein, hydroformylation reactions using ionic liquids, as well as environmental applications such as catalytic removal of pollutants, wastewater treatment using photocatalysis, and converting biomass into fuels and gas. In 2022, she received the Kovalevskaia Prize, a prestigious honor celebrating exceptional contributions by women in science. Mrs. Thang is the HUST coordinator of the RoHan Project.

Le Thanh Son is a Vietnamese chemist with expertise in catalysis and petrochemical chemistry. He earned his Ph.D. from Ehime University in Japan in 1999 after completing his studies at Hanoi University and VNU University of Science. Since 1995, he has been teaching at the VNU University of Science, where he later became Dean of the Faculty of Chemistry and currently serves as a Rector. His research centers on catalyst development for hydrocarbon processing, biodiesel production, and adsorptive materials for pollution control.

Promoting Traditional Medicine, Sustainability, and Internationalisation in Higher Education: The Case of Yaoundé-Bielefeld Bilateral Graduate School of Natural Products with Antiparasital and Antibacterial Activities (YaBiNaPA)

Norbert Sewald, Marcel Frese, Bruno Lenta Ndjakou, and Aloysius Ngefac

Abstract Despite the existence of abundant natural resources in Cameroon, including a very rich biodiversity of flora and fauna, the country is still characterized by a precarious development landscape, plagued by problems such as abject poverty, lack of basic supplies, serious health challenges from bacterial and parasitic diseases, and the unaffordability of synthetic drugs. This precarious development scenario significantly motivated the conception of a South-North collaborative project between the University of Yaoundé 1 (Cameroon) and Bielefeld University (Germany), known as Yaoundé-Bielefeld Bilateral Graduate School of Natural Products with Antiparasite and Antibacterial Activity (YaBiNaPA), funded by the German Academic Exchange Service (DAAD). This chapter assesses this research hub in terms of the way its scientific activities on traditional medicine address the development challenges of Cameroon and those of other developing African nations, promote sustainability, foster inter-university collaboration, and trigger the effective transfer of knowledge. Specifically, the chapter focuses on the goals of YaBiNaPA,

N. Sewald (✉) · M. Frese
Department of Chemistry, Organic and Bioorganic Chemistry, Bielefeld University, Bielefeld, Germany
e-mail: norbert.sewald@uni-bielefeld.de

M. Frese
e-mail: marcel.frese@uni-bielefeld.de

B. Lenta Ndjakou
Department of Chemistry, Higher Teacher Training College, University of Yaoundé I, Yaoundé, Cameroon
e-mail: lentabruno@yahoo.fr

A. Ngefac
Department of English, Higher Teacher Training College, University of Yaoundé I, Yaoundé, Cameroon
e-mail: angefac@yahoo.co.uk

C. Brandt et al. (eds.), *Implementing Sustainability and Internationalization in Higher Education*, World Sustainability Series, https://doi.org/10.1007/978-3-031-97837-1_7

its scientific achievements and knowledge transfer, its efforts to promote the United Nations Sustainable Development Goals (SDGs) and internationalisation in higher education, its networking with health stakeholders, its outreach actions, its efforts to implement the Nagoya Protocol in Cameroon, the challenges encountered, the lessons learned, and the way forward.

1 Introduction

Many developing countries are characterized by a precarious development landscape, whereby basic necessities such as potable water, affordable energy, and basic healthcare are far-fetched dreams, besides the fact that poverty, unemployment, and illiteracy are orders of the day. In the domain of health, the continent of Africa in general and Cameroon in particular are facing enormous challenges, including exposure to many microbial and parasitic diseases and the fact that a large proportion of the population cannot afford primary healthcare and, in most cases, are unable to purchase synthetic drugs, because of the poverty-stricken situation of the continent. This explains why up to 80% of the African population relies on traditional medicine and traditional healers (also known as traditional practitioners) for their primary healthcare.

However, there are many loopholes to traditional medicine or phytomedicine (plant-derived medicine), which need to be addressed before this traditional approach to healthcare can significantly complement modern medicine and provide effective and sustainable solutions to the health problems of the population. The first loophole is the fact that traditional medicine is practised by traditional healers who mostly have no scientific knowledge of the phytodrugs they administer to their patients, especially with regard to their efficacy, active principles, toxicity, dosage, and—above all—storage conditions of plant extracts. In some cases, the phytomedicines are not scientifically validated, their formulations are not standardized, and sometimes there is a problem with the identification and conservation of plant materials. In the Cameroonian situation, traditional medicine concocted from plants is not only of questionable efficacy, but its toxicity is usually unknown and, very often, contributes to very high mortality rate in this African nation. The second loophole is the fact that many plant species (e.g. *Prunus africana*) are seriously threatened and endangered as a result of their over-exploitation, commercialization, chemical contamination and urbanization. The third major loophole is that there are many medicinally rich plant species out there that can provide health solutions to a large proportion of the population, but which have not yet been investigated. In other words, there is lack of scientific investigation on the existence of many herbal plants and there is very little scientific investigation on the toxicity of existing phytodrugs, their dosage and chemical composition. For instance, the dosage prescribed by traditional healers sometimes results in health problems and, occasionally, even in untimely deaths.

With regard to the specific situation of Cameroon, the fact that the country has an extremely rich biodiversity of flora and fauna, but remains a very poor country with chronic health challenges is really paradoxical. This very rich biodiversity can be explored and exploited to combat the numerous parasitic and bacterial diseases affecting the health of the citizens of this country, alleviate poverty and accelerate the development of the country. This was definitely one of the main motivations behind a South-North collaborative initiative between a network of Cameroonian scientists, headed by Professor Bruno Lenta Ndjakou of the University of Yaoundé I (Cameroon), and Bielefeld University scientists, represented by Professor Norbert Sewald. This initiative culminated in the conception of a research hub between the University of Yaoundé I (Cameroon) and Bielefeld University (Germany), known as Yaoundé-Bielefeld Bilateral Graduate School of Natural Products with Antiparasite and Antibacterial Activity (YaBiNaPA), which is financially supported by the German Academic Exchange Service (DAAD), with funds from the German Federal Ministry for Economic Cooperation and Development.

Considering that parasitic and microbial infections are alarming public health challenges in Cameroon and in developing African countries and given that medicinal plants of different species abound in the rich biodiversity of tropical African countries, including Cameroon, YaBiNaPA was conceived as a research hub that would subject traditional medicine (phytomedicine) into scientific investigation, provide scientific knowledge to traditional healers who rely on herbal plants to provide healthcare to a significant proportion of the population, and develop a phytodrug industry in Cameroon for economic, health, and scientific benefits and for the sustainable development of developing African countries. In other words, this South-North collaborative initiative concentrates its scientific activities on traditional medicine to address the development challenges of developing African nations in general and those of Cameroon in particular, promote the United Nations Sustainable Development Goals (SDG), foster internationalization in higher education, and trigger the effective transfer of knowledge.

This article is therefore an assessment of the activities of YaBiNaPA in Cameroon and beyond in the context of traditional medicine (phytopharmacology and phytomedicine) and focuses specifically on the following: (a) YaBiNaPA and its goals, (b) the scientific achievements of YaBiNaPA and knowledge transfer, (c) YaBiNaPA and the promotion of the United Nations SDG, (d) YaBiNaPA and the promotion of international collaboration in higher education, (e) YaBiNaPA and networking with health stakeholders, (f) YaBiNaPA and Outreach, (g) YaBiNaPA and the implementation of the Nagoya Protocol in Cameroon, (h) Challenges encountered by YaBiNaPA and its efforts to overcome them, (i) YaBiNaPA and the lessons learned from the project, and (j) YaBiNaPA and the way forward.

2 Background: Natural Products as Medicinal Drugs and the Situation of Africa and Cameroon

Before making an appraisal of the activities of YaBiNaPA with regard to the aspects enumerated above, it is worthwhile to focus first of all on the use of natural products as medicinal drugs and how they are used in Africa and Cameroon by traditional healers (also known as herbalists or traditional practitioners) to provide primary healthcare to a significant proportion of the population, besides exploring the industry for economic reasons.

Natural Products as Drugs

The importance of traditional knowledge for drug development has been highlighted by Tu Youyou, the 2015 Nobel Laureate in Physiology and Medicine. After discovering that Traditional Chinese Medicine can rely on sweet wormwood (*Artemisia annua*) to treat fever, she received the award for being the main architect for the discovery of a novel therapy against malaria. Natural products from plants or microorganisms form the basis of innovative therapeutic agents. About 35% of modern medicines is derived directly or indirectly from medicinal plants (Butler 2004). Recent progress in the search for drugs from plant sources has provided new and important lead compounds against various pharmacological targets, including malaria, cancer, and pain. For example, two of the most important antimalaria drugs, quinine and artemisinin, were isolated from the medicinal plants *Cinchona officinalis* and *Artemisia annua*, respectively (Veldsema and Van Geldre 1973 and Joyce 1993). Taxol (Paclitaxel) is a clinically used anticancer agent isolated from plants. Moreover, many antibiotic compounds or, at least, their parent compounds (e.g. penicillins, cephalosporins, vancomycin) are natural products (Balunas and Kinghorn 2005). Therefore, there is a high potential that there are many active compounds still to be discovered that can be isolated in an ethnopharmacologically driven screen. Standardized extracts containing such compounds can be very useful to those who cannot afford synthetic drugs. This implies that therapeutic approaches based on ethnopharmacological drugs and guidelines also require standardized herbal medicinal formulations. Medicinal plants are therefore bio-resources with high cultural and economic potentials that can contribute sustainably to the economic emergence of developing countries, if they are subjected to thorough scientific investigations.

The Use of Traditional Medicine in Africa

Compared to Traditional Chinese Medicine, African Traditional Medicine is still underexplored in a scientific sense. In Africa, a large percentage of the population (up to 90%) uses traditional medicine as primary healthcare (Parasuraman 2018). Traditional medicine has maintained its popularity in all regions of the continent and its use is rapidly spreading to industrialized countries. Although Western medicine is generally accepted throughout Africa, it has rather complemented indigenous healthcare, instead of replacing it. This can be attributed to cultural preferences as well as to the high cost and the unaffordability of Western medicine. It is also worth pointing out that tropical African countries are sites for rich resources of traditional remedies, with the availability, accessibility and affordability of traditional practitioners. The gap between modern and traditional practitioners has therefore been significantly narrowed during the past decade (WHO fact sheet 1996).

In Africa, many plants derived products are used by traditional healers for the treatment of diseases. It has been estimated that approximately 80% of the population in developing countries relies on traditional medicine for their primary healthcare. A high percentage of the population is poor and living in rural areas, close to natural resources and has rich traditions that depend on medicinal plants. However, most of the species used have never been investigated with regard to efficacy, active principles, toxicity and—above all—the conditions of storage of the formulations. The training of qualified professionals in the valorization of the available bio-resources becomes necessary for the improvement of their health status, the reduction of poverty, and the enhancement of economic development in a sustainable way.

Traditional Medicine and Traditional Healers in Cameroon

There is a continuous need for new and improved drugs to fight parasitic and microbial diseases, which still remain serious public health problems in tropical and subtropical regions of the world. The Congo Basin in general and Cameroon in particular have a rich biodiversity, comprising about 90% of the African ecosystem, which is made up of diverse ethnic groups, each constituting a unique ethnopharmacopoeia that makes Cameroon the richest therapeutic patrimony in Africa. With more than 8260 plant species and several groups of animals, Cameroon has a significant diversity of flora and fauna, ranking 5th in Africa (Letouzey 1985 and Onana 2015). Plants grow everywhere in Cameroon and are used differently according to the different Cameroonian cultures. This explains why the country is a source of drugs and a potential source of income for its poverty-stricken population and for the world in general. But insufficient documentation, lack of scientific experimentation for the verification of the healer's claims, the over-exploitation and commercialization of species, like *Prunus africana*, and the chemical contamination and urbanization that are significantly jeopardising natural habitats for herbs serve as threats to the sustainability of traditional medicine and drug discovery from plants. In Cameroon, about 115 plant species used in traditional medicine are classified as threatened. The sector of traditional medicine and medicinal plants exploitation is not well organized because government/policy makers and financial investors do not pay much attention to herbal medicine as a means of creating wealth and transforming lives, especially for the poor people (Osunderu 2009). There is lack of scientific investigation on the toxicity, dosage and chemical composition of phytodrugs. The doses given by the traditional healers might result in health problems and cause side effects that can be deadly.

Medicinal plants are bio-resources with high cultural and economic potentials that can efficiently and tremendously contribute to the sustainable and economic emergence of developing countries, if they are scientifically investigated. Many institutions have an interest in studying and securing the future of medicinal plants, especially because of the direct benefits they provide to the local population and their importance in the wider context of biodiversity conservation. It is important to focus on the role of medicinal plants in healthcare and development, in biodiversity conservation, in the use of traditional knowledge, intellectual property rights and in benefits sharing, marketing and trade.

Research groups in Cameroon and all over Africa are actively engaged in an ethnobotanical/guided search for active principles from medicinal plants used in traditional medicine. The aim has been to extract, isolate, and structurally characterize plant metabolites, which could be developed into modern therapeutic agents. Research on medicinal plants from Cameroon and their endophytes has led to the isolation of more than 10,000 secondary metabolites documented with more than 4500 publications in international peer-reviewed journals. However, these studies conducted for several decades have often been limited to preliminary screening or determination of the chemical composition and have not consequently produced satisfactory results. In a similar way, the formulation of standardized herbal medicines or the scientific validation and rationalization for using the species used in traditional medicine are still underdeveloped. In addition, some over-exploited species are threatened with extinction. This is attributed to the lack of (a) **expertise** on phytomedicine formulation, (b) **platforms for testing** and (c) **scientific networks** between traditional healers, ethnobotanists, plant taxonomists, phytochemists, biochemists, and pharmacologists. Even in cases where secondary metabolites displayed promising activity on some strains of microbes or high toxicity, the feedback is not always given to traditional healers.

In Cameroon, many plants are used for the treatment of microbial and parasitic diseases. The sale of medicinal plants takes place not only in phytotherapeutic centers but also in open markets and even as mobile business on inter-regional Express Buses. In 2015, a survey done in different regions of Cameroon reported that 77.4% of the population takes traditional medicines at least once per year (Abia et al. 2015). More than 90% of the Cameroonian population (like the Sub-Saharan population in general) depends on non-standardized traditional medicinal plant concoctions of questionable efficacy and unknown toxicity and this contributes to the extremely high mortality rate in this African region.

3 YaBiNaPA as a Research Hub and its Goals

Prior to the conception of YaBiNaPA, there was a network of Cameroonian scientists who had been collaborating with the group of Norbert Sewald at Bielefeld University. The Cameroonian coordinator, Bruno Lenta Ndjakou, had been visiting Bielefeld University as a Marie Curie Postdoctoral Fellow for two years and in 2013 he was able to obtain a Georg Forster Fellowship for Experienced Researchers. During this second fellowship stay at Bielefeld University, the conception of YaBiNaPA became a reality in 2016.

It aimed to develop the African pharmacopoeia, which would serve as a Cameroonian research hub and exploit the rich biodiversity of the country, educate and orientate traditional practitioners who unsustainably use this rich biodiversity of flora and fauna, and provide the scientific knowledge necessary for the sustainability of medicinal drugs and for the promotion of the United Nations SDG. Financial Support came from the German Academic Exchange Service (DAAD), with funds from the German Federal Ministry for Economic Cooperation and Development. YaBiNaPA laboratories have been equipped with state-of-the-art equipment for natural products chemistry and for biochemical/biological assays. A laboratory platform for chemical separation (preparative medium pressure liquid chromatography, MPLC) and analysis (liquid chromatography–mass spectrometry, LC–MS) has been established. In addition to the analysis platform, further state-of-the-art equipment (a second low-resolution Tandem quadrupole LC–MS/MS system, a GC–MS and a GC–MS/MS system) was obtained through the collaboration with the Recycling Organization for Research Opportunities (RORO, UK). It is worth mentioning that a new biosafety level 2 cabinet can now be used to analyse malaria and leishmania parasites for the elucidation of the biological activity of the extracts and isolated compounds. Further structure analysis with nuclear magnetic resonance spectroscopy (NMR) and X-ray single-crystal structure analysis are done at Bielefeld University. A campus-wide license for SciFinder (a research discovery application that provides integrated access to the world's most comprehensive and authoritative source of references and substances in chemistry and related sciences) has been acquired and can be accessed by all YaBiNaPA fellows. In addition, a computer room for literature research and manuscript preparation has been newly equipped.

Following the WHO Traditional Medicine Strategy 2014–2023, YaBiNaPA embarked on the following goals:

- To improve the training in graduate education using a network of experts in natural products research and modern research facilities, which provide a broad scope of translational and interdisciplinary skills;
- To harness the potential contribution of traditional medicine (phytomedicine) to good health, well-being and people-centered healthcare;
- To promote the safety, efficacy and quality of phytomedicine by expanding the bases and scope of scientific knowledge;
- To generate guidelines for traditional medicine by developing and providing international standards, technical guidelines and a valid quality control for formulations and methodologies. This is achieved through a continuous exchange between traditional healers and researchers at eye level, combined with quality education and trainings;
- To increase the sustainable availability and affordability of traditional medicine, with emphasis on its accessibility to the poor and suffering population; and
- To create a platform for exchange with traditional healers on the conservation and sustainable use of medicinal plants.

YaBiNaPA therefore provides the basis and facilities for learning, knowledge transfer, and multidisciplinary research and analysis in ethnobotany, ethnopharmacology, phytochemistry, biochemistry, and drug development. To be specific, the formulation and quality control of herbal phytomedicine are important prerogatives for YaBiNaPA. A multidisciplinary structured training providing knowledge transfer and scientific exchange through lectures, seminars, conferences, publications and workshops, depending on the target group, has been organized by YaBiNaPA. For Ph.D. students, the training program relies on concerted learning, a combination of introductory training, hands-on learning "on the bench", teaching by peers, and training in additional skills. They explore and validate methods for the development of extracts or compounds for phytomedicine, while investigating the molecular mechanisms in microorganisms. In this light, a network of experts from Cameroonian universities, other African universities, universities in Europe and private companies has given out several modules and capacity building courses to students both face-to-face and online. Some of the modules and capacity building courses includes:

- A theoretical and practical training course on "the use of the HPLC and LC–MS/MS", given by Dr. Giles Edwards, University of Manchester, UK
- Theoretical and practical training course on Analytical Chemistry and HPLC-analysis, given by Dr. Olivier Heudi from the Novartis Institutes for Biomedical Research, Switzerland
- How to write a good paper—webinar series on scientific writing, by Dr. Kerrin Riewerts (Bielefeld University)
- A training course on entrepreneurship, by Dr. Charles Mboning.

In terms of teaching and research, the University of Yaoundé I benefits greatly from the YaBiNaPA project. Platforms equipped with state-of-the-art equipment for chemical and biological analyses have been built at the University of Yaoundé I for the training of experts and for the valorization of plants. In addition to laboratory equipment and the provision of reagents and consumables to students for their work, several other non-monetary benefits (such as publications, researchers' mobility, the University of Yaoundé I connection to the databases of the hub, and capacity building courses offered by international experts) have made YaBiNaPA a pole of excellence for the valorization of plants and ancestral knowledge, for the improvement of the wellbeing of the population, and for sustainable development.

The Scientific Achievements of YaBiNaPA and Knowledge Transfer

YaBiNaPA has developed one of the best platforms for chemical and biological analysis of plants in the sub-region. In fact, the ongoing research projects under YaBiNaPA focus on three main areas:

(1) Search for antibacterial and antifungal agents;
(2) Search for potent antiparasitic agents from plants; and
(3) Formulation and quality control of phytodrugs.

With the aim to train experts in the sustainable development of plants, the transversal research approaches of YaBiNaPA make it possible to engage several students from different disciplines (e.g. chemistry, biochemistry, and biology) on specific projects for the attainment of common objectives. This multidisciplinary structuring and training facilitate knowledge transfer.

With regard to the scientific achievements of the hub, it is worth reporting that, through consultations of databases and with the help of an ethnopharmacological survey conducted in different localities of Cameroon, including Bandounga and the pygmies community of the Eastern Region of Cameroon, more than 250 plants were selected for scientific experimentation. After harvesting, chopping, drying and crushing, the different parts of these plants were extracted with organic solvents and hydroalcoholic mixtures. More than 900 extracts were prepared. All the extracts were screened for their antimicrobial activities against 22 strains of bacteria and fungi and for their antiparasite activity against *Plasmodium falciparum* (D2 and 3D7 strains) and *Leishmania donovani*. Concerning antiparasitic activities, more than 85 extracts were active with more than 38 extracts showing promising activity.

As concerns antimicrobial activities, more than 550 extracts were active at least on one tested strain. All active extracts were selected for a bio-guided investigation, with the aim to isolate the active constituents. The species already investigated under YaBiNaPA led to the isolation of more than 2000 secondary metabolites, including about 140 new compounds from different classes (e.g. alkaloids, xanthones, flavonoids, steroids, norbergenins, terpenoids, lignans, limonoids). Active and non-toxic extracts have also been selected and are in the process of formulation. Syrups against salmonellosis and malaria were prepared from the hydroalcoholic extract of plants (*D. microcarpum*, *Nauclea latifolia*, and *Dacroydes edulis*). In addition, a soap cream against fungal skin infections was prepared from the leaves of *C. alata*. These improved traditional medicines will be subjected to further clinical testing.

YaBiNaPA and the Promotion of the United Nations SDGs

In addition to the scientific achievements of YaBiNaPA, the hub's activities significantly promote the United Nations Sustainable Development Goals (SDGs), especially SDG 1 (No Poverty), SDG 3 (Good Health and Wellbeing), SDG 4 (Quality Education), SDG 8 (Decent Work and Economic Growth), SDG 10 (Reduced Inequalities), and SDG 15 (Life on Land). This section discusses the role of YaBiNaPA in the promotion of the United Nations Sustainable Development Goals.

With regard to SDG 1 (No Poverty), the traditional medicine industry in Cameroon and in other African countries is a major source of income to traditional healers who provide primary healthcare to a very large proportion of the population. Apart from the fact that bio-resources serving as medicinal plants are on a high demand and significantly alleviate poverty to those who cultivate them, traditional healers rely on them to provide healthcare to the population. But the success of traditional healers in the traditional medicine business depends on the quality of health services they are able to provide. This explains why YaBiNaPA should be one of the key stakeholders in this industry, as the hub provides scientific knowledge to traditional healers and ensures the sustainability of medicinal plants. This means the survival of this industry, a source of poverty alleviation to traditional healthcare providers, significantly depends on the scientific knowledge and orientation YaBiNaPA provides.

In other words, YaBiNaPA does not only provide scientific knowledge to traditional healers, the hub guides against over-exploitation of bio-resources and other malpractices that can jeopardize the sustainability of medicinal plants and compromise the traditional medicine industry as a source of poverty reduction. It can therefore be maintained that herbal plants and phytomedicine (promoted through YaBiNaPA activities) provide cultivators and traditional healers additional income and contribute to poverty reduction, addressing issues such as hunger and insecurity, while enhancing livelihoods in rural areas. The scientific approach offered by YaBiNaPA actually enhances the sustainability of medicinal plants and the traditional medicine industry.

As concerns SDG 3 (Good Health and Well-being), YaBiNaPA is largely about good health and well-being through phytomedicine. Considering that parasitic and microbial infections are alarming public health challenges in tropical countries such as Cameroon and given the evolution of bacterial and parasitic diseases, the resistance of microorganisms to existing drugs, and the inability of a large proportion of the population to afford synthetic drugs, traditional medicine and traditional healers are unavoidable in providing healthcare to the population. Plants from different families are used by traditional healers to treat microbial and parasitic diseases (Phytomedicine). For many years, some of these plants have been the focus of many scientific investigations in Cameroon and Africa, but most of these investigations simply produced knowledge on the species investigated, without adequate scientific knowledge on the medicinal properties of the plants. This explains why one of YaBiNaPA's missions is to develop strategies to go from plants to efficient, nontoxic and affordable phytodrugs, as a concrete attempt to promote the United Nations SDG 3. After all, the World Health Organisation (WHO) advocates the incorporation of safe and effective traditional medicine into primary healthcare systems (Akinyemi et al. 2018).

Regarding SDG 4 (Quality Education), the activities of YaBiNaPA have had an enormous impact on the training of junior scientists and young scholars, especially with regard to teaching, research methodology and capacity building. Many scientists from the University of Yaoundé I and from other Cameroonian universities have lauded the scientific activities of the hub and the impact it is creating on the quality of training in higher education in Cameroon and beyond.

As concerns SDG 8 (Decent Work and Economic Growth), the activities of YaBiNaPA significantly promote phytomedicine and economic growth. Several African countries have projected their economic emergence to take place in 2030, but this cannot be possible without well-trained and healthy human resources. Therefore, the efforts of YaBiNaPA to improve the health status and the health quality of the population in a sustainable and affordable way have a great impact on economic development. A safe approach towards the use of plant-derived medicine (Phytomedicine) requires well-trained researchers and well-educated healthcare stakeholders such as traditional healers who receive scientific knowledge and orientation from YaBiNaPA scientists.

Concerning SDG 5 (Gender Equality) and SDG 10 (Reduced Inequalities), YaBiNaPA does not discriminate on the basis of gender and does not promote any form of inequalities in its scientific activities, in the teaching and training of junior scientists and in the amelioration of traditional healthcare to the population. Considering that the female gender is underrepresented in Cameroonian science, like in many other countries, females are particularly encouraged through the activities of YaBiNaPA, and are given equal opportunities like their male counterparts. YaBiNaPA also strives to avoid any form of inequalities by providing equal opportunities to everyone, irrespective of their religious beliefs, ethnic background, political stance and even nationality. This explains why beneficiaries of research and training opportunities offered by YaBiNaPA are not only Cameroonians. For instance, besides the 36 Cameroonian Ph.D. students selected to travel to Bielefeld under the auspices of YaBiNaPA research hub, up to 120 university students from other African countries (e.g. Democratic Republic of Congo, Benin, Gabon, Madagascar, Niger, and Tchad) have been selected to reside either in Yaounde or Bielefeld for research reasons. This is a practical way by YaBiNaPA to avoid any form of discrimination and to reduce inequalities.

Regarding Life on Land (SDG 15), YaBiNaPA acknowledges and maintains that medicinal plants are sources of potent therapeutic agents. The impact of natural products in the discovery of new lead drugs is due to their amazing structural diversity, which makes them a seemingly infinite source of chemical compounds (Shu 1998 and Calixto 2019). Hence, medicinal plants are important resources that can contribute to the economic development of tropical countries. It is thus evident that only a sustainable use of these bio-resources can contribute to the improvement of the health of the population, the reduction of poverty and the valorization of ancestral knowledge. Therefore, the development and application of phytomedicine must be intimately connected with the protection of the natural resources found on the surface of the earth.

YaBiNaPA and the Promotion of International Collaboration in Higher Education

The first phase of the project ended in 2020. During this phase, 58 Ph.D. students from Cameroonian Universities (Bamenda, Buea, Douala, Dschang, Maroua, and Ngaoundéré), 13 students from other African universities (Congo, Benin, Ethiopia, Gabon, Madagascar, Malawi, Niger, Uganda, Tchad, Gabon and Zimbabwe) and 6 postdocs were selected for research stays in Bielefeld or Yaoundé. In addition, professors and researchers from Cameroon had 10 research stays in Bielefeld. Amongst the 58 Ph.D. stays for students from Cameroonian Universities, 20 were selected for full 4 year-Ph.D. programmes. The duration of the other stays varied from 3 to 12 months. Given the impact of YaBiNaPA's activities between 2017 and 2020 on teaching, research methodology and the capacity building of scientists at the University of Yaoundé I and other Cameroonian universities, several researchers from other Cameroonian universities found the programme innovative and have joined the network for the second phase of the project. In 2021, the number of PIs was then extended to 27.

Since 2021, 36 Ph.D. students from Cameroonian Universities, 120 students from other African universities (Democratic Republic of Congo, Benin, Gabon, Madagascar, Niger, Tchad) and 4 postdocs were selected for research stays at Bielefeld University and at the University of Yaoundé 1. Furthermore, more than 10 professors from Cameroon have conducted research at Bielefeld University during different research stays. The mobility programmes led to the creation of a strong network between YaBiNaPA and other African universities, including the supervisors of the students selected in the project. Moreover, guest professors from Turkey, France, England, Belgium, Ghana and other countries gave lectures to students, as a further attempt to enhance their capacity.

YaBiNaPA and Networking with Health Stakeholders

It is the declared goal of YaBiNaPA to establish an **Excellence Centre for Natural Products Research and Phytomedicine** that includes scientists working in other universities in Cameroon and beyond for the development of new concepts for plant-derived therapies. Hence, YaBiNaPA has been very active in national and international exchanges between researchers in Cameroon and those from other African countries and Germany.

The research network has been expanded by establishing a multitude of contacts, not only between YaBiNaPA and other Cameroonian universities in Cameroon (e.g. University of Bamenda, University Buea, University of Douala, University of Dschang, University of Maroua and University of Ngaoundéré), but also with local authorities, companies, clinicians, traditional healers and the civil society. Between 2016 and 2024, numerous agreements on mutual collaborations have been signed between YaBiNaPA/University of Yaoundé I and the following institutions in the academia, the public sector, and the private sector:

- Centre Pasteur (CPC), Cameroon;
- Institute of Medical Research and Studies of Medicinal Plants (IMPM), Cameroon;
- National drugs Quality Control and Validation Laboratory (LANACOME), Cameroon;
- Phytotherapy Centre Plante Sante and Vie, Cameroon;
- Alango Foundation, Center of Phytotherapy, Dschang, Cameroon;
- Molecular Biology Center, Yaoundé, Cameroon.
- Recycling Organization for Research Opportunities (RORO).

Joined projects have been conceived with the following phytotherapy centers: **Polybucal**, **Fondation Gefeh** and **Groupe Foba** for the investigation of their formulation and quality control. Collaborations have been initiated with private companies such as **Laboratoire Roger Ducos**, which produce traditional medicine products and cosmetic products from plants.

At the international level, an **International Scientific Advisory Board** has been set up, including Professor Catherine Vonthron-Senecheau (Strasbourg/France),

Professor Kerstin Andrae-Marobela (Gaborone/Botswana), Professor Philip Rosenthal, (San Francisco/USA), Professor Samuel Asare-Nkansah (KNUST, Kumasi/Ghana), Prof. Dr. Syed Ghulam Musharraf (Karachi/Pakistan) and Professor Ossy MJ Kasilo (WHO Regional Adviser for Traditional Medicine, Dar-es-Salaam/Tanzania).

YaBiNaPA is involved in a worldwide network with foreign scientific partners. It has agreed to collaborate with two African Centres of Excellence: the Malawi-based ***African Centre of Excellence in Public Health and Herbal Medicine*** (**ACEPHEM**, Dr. Fanuel Aaron Lampiao, Professor Adamson Muula) and the Uganda-based ***African Centre of Excellence in Pharm-Bio Technology and Traditional Medicine*** (**Pharmbiotrac**, Dr. Patrick Engeu Ogwang).

With regard to traditional healers, YaBiNaPA regularly organises workshops to foster collaboration with them, including local authorities, pygmy communities, members of Traditional Healers Association of Cameroon, and the Baka pygmy community of East Cameroon. In 2019, experienced YaBiNaPA members organised many workshops on traditional medicine, scientific research, collaborations for sustainable development, ethnopharmacological survey, and theory and practice in traditional medicine research. During these meetings, several contacts with traditional healers were established, resulting in a much more fruitful collaboration between traditional healers and YaBiNaPA researchers. In addition, YaBiNaPA also organised workshops to train collaborators, researchers and non-YaBiNaPA students from the University of Yaoundé 1 and other Cameroonian universities on the use of the HPLC, MPLC, LC–MS systems and how to carry out antimicrobial susceptibility testing.

Traditional healers play a key role in promoting ancestral knowledge. In tropical Africa, they are often the first and last line of defense against the most contagious and debilitating diseases that plague lives. Traditional medicine has maintained its popularity in all regions of the developing world and its use is rapidly spreading to industrialized countries. This system of medicine is reserved for healers who are influential to the community and are often taken more seriously than any other specialist of health. In Cameroon, many communities have conserved the bio-resources used for traditional medicine by relying on their indigenous skills. Phytotherapy centers and traditional healers are present in all the cities and villages of the country. Some communities such as **Baka and Bagyeli pygmies** live in the forest and represent one of the minority population groups of the country to use only traditional medicine for their primary healthcare. They are known by all Cameroonians as the best healers in the country. Even people from other parts of the country and abroad come to receive medical care from them. Pygmies, respecting their traditions, use the resources made available to them by nature to heal themselves. They make their own remedies, partly because they are attached to a way of life inherited from their ancestors and partly because, without money and without administrative papers, they do not have access to the care provided by the hospital.

This explains why one of the future missions of YaBiNaPA will be to establish a **training and information center for traditional healers, who provide primary healthcare to many members of the pygmy community**. The evaluation of the efficiency, toxicity, formulation, and quality control of the formulations of traditional healers is necessary for a safe contribution to a sustainable health system. Thus, the collaboration between traditional healers and scientists in Cameroon is important for the scientific validation of formulations and, above all, for the conservation of active plant species for sustainable development. Interactions with traditional healers need to ameliorate, given that mutual trust and confidence are indispensable elements in the traditional medicine industry. Besides, a fruitful interaction with traditional healers must rely on transparency, data sharing, and joint decision-making. In response to these preconditions, YaBiNaPA has already established strong collaborative ties with traditional healers of the Center Region and the pygmy community of the Eastern Region of Cameroon. These ties are being strengthened through yearly trainings YaBiNaPA offers to these traditional healers. This significantly facilitates the selection of species used in the treatment of parasitic and microbial diseases. In addition, traditional healers receive information on the results obtained from the investigations conducted by YaBiNaPA scientists.

YaBiNaPA and Outreach

As part of YaBiNaPA's research communication strategy, a video about YaBiNaPA was produced in 2019 by Research TV of Bielefeld University and was published and disseminated on numerous social media platforms, such as YouTube, Facebook, the Bielefeld University Blog and Twitter. Another one is to be released in the nearest future. The important results, events, and news of YaBiNaPA were also disseminated on the biannual newspaper *YaBiNaPA Actu*. Open days were organized and were valuable opportunities for scientists to inform the public about their works, highlight their knowledge, and, above all, meet researchers from other institutions, traditional healers, political decision-makers, and people from the public and private sectors.

An open day of YaBiNaPA was also organized on the campus of the Higher Teacher Training College (University of Yaoundé I) on the 24th of January, 2020. Attendees included more than 150 participants, including representatives from the administration (University of Yaoundé I, Ministry of Scientific Research, and Ministry of Public Health), community leaders, students, scientists, traditional healers, and members of the public and private sectors. The activities started with a **panel discussion** on the topic ***"Medicinal Plants and Sustainable Development"***. The panel was made of persons with long-standing experience in the use of medicinal plants, research, and legislation in the field of medicinal plants exploitation. Aspects such as the history of traditional medicine, biodiversity in Cameroon, ethnopharmacology, sustainable use and conservation of species, contribution of phytomedicine to healthcare and economic development, and the necessity for quality control of plant-derived formulations were developed during the lively and fruitful discussions. Many participants recommended the re-organisation of similar scientific events in the future. After the panel discussion, the participants had the opportunity to interact with YaBiNaPA students who displayed products formulated from plant extracts (Fig. 1). In addition, it was an opportunity to establish contacts with healers, heads of communities, people from the private sector, and heads of phytotherapy centers. The paradigm shift brought about by YaBiNaPA in the study of plants was highly appreciated by State authorities who promised to contribute significantly to the project

Antifungal cream Antisalmonella syrup Antifungal soap

Fig. 1 Exhibition of products prepared by YaBiNaPA students during the open day. Additional pictures can be found at www.yabinapa.de

of restructuring and revamping the industry of traditional medicine and medicinal plants in Cameroon.

Considering that the dissemination of scientific results is very important, YaBiNaPA students and instructors have been sharing their scientific findings with other scientists through international conferences, weekly seminars, workshops, and Ph.D. research works.

YaBiNaPA and the Implementation of the Nagoya Protocol in Cameroon

YaBiNaPA has been implementing the Nagoya Protocol on Access and Benefit Sharing (ABS). Each project in the framework of YaBiNaPA follows the regulations of the Nagoya Protocol. This makes YaBiNaPA a pioneer in terms of the implementation of the Nagoya Protocol in the academia in Cameroon. In addition, YaBiNaPA has been working in collaboration with communities and policy makers to ensure the sharing of benefits in the case of a breakthrough from its research activities. During field studies and meetings with communities, the results of the investigations of the species are presented. Thus, the species with a high potential of valorization are highlighted as well as those with high toxicities. Furthermore, the skills of members of the communities are also reinforced on the methods of sustainable harvest, culture, and the conservation of plants and derived products for the improvement of the life of the populations. The creation of collaborative platforms between YaBiNaPA and traditional healers facilitates access to scientific knowledge and data from plants research.

It is evident that, through its various activities, YaBiNaPA contributes to the implementation of the Nagoya Protocol in Cameroon. Capacity building courses for students and other members of YaBiNaPA on the protocol have been provided by the focal point of the Nagoya Protocol in Cameroon. The creation of databases on ancestral knowledge and the collection of data from scientific studies carried out on plants will allow a better classification of resources. In order to comply with the research expectations of the Nagoya Protocol, YaBiNaPA has organized a panel discussion on the theme "Interest of identification of resources with high potential of valorization for a better implementation of the Nagoya Protocol in Cameroon". The participants included representatives from the administration (University of Yaoundé I, Ministry of Scientific Research, Ministry of Public Health, Ministry of Environment, Nature Protection and Sustainable Development), community leaders, students, scientists, traditional healers, and members of the public and private sectors.

In the case of YaBiNaPA, a framework material transfer agreement had been concluded between Bielefeld University and the University of Yaoundé 1, stating that the inherent intellectual property rights and proprietary information regarding the genetic material shall remain the exclusive property of the Cameroonian Provider. The plant material for the first funding period (2016–2020) had been collected before Cameroon became Party of the Nagoya Protocol (28.02.2017) and could thus be used by YaBiNaPA without any restrictions. All genetic resources collected for the second funding period were subject to the regulations of the Nagoya Protocol. The Cameroonian ABS Procedure distinguishes four different access types. Bioprospecting research and the use of genetic information by professionals belonging to the national research system (access type 3) requires relatively little paperwork. However, as soon as the genetic material is transferred outside the country by a national researcher—even without any commercial purposes—access type 1 requires issuing of a PIC (Prior Informed Consent) and negotiation of a MAT (Mutually Agreed Terms). This is also applicable to the investigations that are being carried out at Bielefeld University, despite the fact that all rights are being retained by the Cameroonian researchers. We consider this a major obstacle for Cameroonian scientists in general, when it comes to collaborations with foreign academic institutions.

Although the project is still ongoing, it is not too early to state that YaBiNaPA has been successful in implementing a North–South partnership with a strong reference to the UN SDGs and the Nagoya Protocol on Access and Benefit Sharing. A strong scientific network in Yaoundé, in Cameroon, and with many researchers and institutions within Africa has been established. Emphasis continues to be placed on graduate student education and contributions to a Cameroonian Pharmacopoeia.

4 Challenges Encountered by YaBiNaPA and its Efforts to Overcome Them

In spite of the interesting success stories of YaBiNaPA shared so far in this chapter, the research hub has encountered some challenges, which include gender-related stereotypes that tend to discriminate against women, the unstable political atmosphere in Cameroon, the unsustainable practices of traditional healers, cultural exigencies, and unnecessary bureaucratic hurdles from customs. These challenges will be discussed, with a focus on the way they were overcome.

Gender-Related Stereotypes that Discriminate Against Women

With regard to gender-related stereotypes and cultural beliefs that promote inequalities between men and women and discriminate against the latter, we consider this as a challenge to YaBiNaPA. It is worth pointing out that, because of cultural ideologies, especially the tendency in the African context to consider science as a difficult area reserved mainly for men, women working in scientific fields are victims of social discrimination (Breda 2014). Gender studies in Cameroon show that the absence of women in scientific professions is the consequence of a biased social construct, which claims that certain professions are not meant for women and constitute the exclusive domain for men.

According to data from the University of Yaoundé I, women enrolled in scientific disciplines very often obtain very good results and this clearly underscores their aptitude—beyond any socio-cultural considerations. Although the gap between the genders has been narrowed during the past 10 years in most science degree programs up to the Ph.D., women are less likely than men to opt for a scientific career after the Ph.D. phase, because of multiple reasons. In 2019, among the 320 lecturers of the Faculty of Science at the University of Yaoundé I, 23% were women. Moreover, women are poorly represented on decision-making fora in the different universities in Cameroon. Few of them coordinate research groups or are head of laboratories. YaBiNaPA is striving to integrate female scientists in the team of principal investigators. In the sustainable development approach, it is necessary to adopt strategies to encourage more female scientists to remain in the academia. Potential obstacles obstructing women from pursuing a career in science need to be removed.

Hence, gender equality is one key element in the recruitment strategy of YaBiNaPA within the broader context of diversity management approach. Human diversity is a desirable asset in itself for a programme that strives not only to gain new knowledge, but also to provide education in a broader sense to young scientists. Female principal investigators will become role models in YaBiNaPA. The idea is to let female students interact with female scientists to gain inspiration and motivation. Research data show that having a female supervisor improves both the level of female students in these subjects and their desire to pursue scientific studies.

In the first funding period, 8 out of 20 full time Ph.D. students were female, whereas female principal investigators were largely underrepresented. The latter fact was highly unsatisfactory and YaBiNaPA succeeded in integrating 8 additional female principal investigators for the proposed second funding period. YaBiNaPA intends to have a female share of at least 50% regarding funded projects for doctoral students and postdocs. For the second phase of the research hub (2021–2025), 50% of the Ph.D. students selected from the University of Yaoundé 1 for a full Ph.D. program are women. This was successful because of the gender-sensitive calls for application, additional announcement channels exploited and the proactive desire to give female candidates the opportunity to valorise their potentials. YaBiNaPA also strives to address and consider the different life patterns, needs and interests of both women and men. In addition, laboratory work is not imposed on expectant and nursing mothers. Therefore, female scientists expecting childbirth are entitled to receive support for home offices and temporary technical assistance in the laboratory.

The Unstable Political Atmosphere in Cameroon and in Neighbouring Countries

Cameroon currently suffers from two violent conflicts between the government and separatists from the English-speaking minority (which is now known as the Anglophone Crisis) and a renewed Boko Haram insurgency in the Far North Region. These conflicts, together with the political unrest in the Central African Republic, have led to the displacement of nearly 1 million people in Cameroon and constitute a major challenge to the country, but also to YaBiNaPA. These political conflicts make research very difficult in the affected regions, which host three Cameroonian universities (i.e. North West Region (University of Bamenda); South West Region (University of Buea); and Far North Region (University of Maroua)). Cameroonian scientists in these three universities are seriously affected by the conflicts and deserve additional support. Considering that some of the scientists in the affected regions are research partners of YaBiNaPA, the ongoing conflicts in the three regions directly affect the hub. This explains why YaBiNaPA prioritizes the integration of scientists from these regions into the activities of the hub and strives to provide them all the necessary support.

The political circumstances affected YaBiNaPA's performance, too. Because of the unstable situation in the regions North-West and South-West, solvents and other petrochemical products were not available in Cameroon for longer periods of time, which strongly compromised research performance.

The Potentially Lacking Sustainability of Traditional Healers

In fact, traditional healers are not often aware that the increasing commercialization of plants has resulted in the overharvesting of some economically important medicinal plant species, many of which have become rare, endangered and threatened. More than 630 species in Cameroon are listed as threatened on the International Union for Conservation of Nature (IUCN) Red List, of which 183 are endangered and 115 are critically endangered. This is the case of *Prunus africana* that is threatened by extinction due to commercial overexploitation (Ingram et al. 2015 and Mpouam et al. 2022). In Cameroon, *P. africana* is a special non-timber forest product (NTFP) of interest. Cameroon was the world's largest exporter of *P. africana bark* in 2012 (Awono et al. 2016 and Tassiamba et al. 2022). In fact, its bark is the basic material for the industrial manufacture of medicines against prostate disorders. Since the 1990s, the demand for *P. africana* bark in the world has been increasing. The world demand is estimated at 4000 tons per year for a value of finished products estimated at 220 million US dollars. The survival of the species in its natural wild habitat is threatened by unsustainable and uncontrolled exploitation.

Growing the tree in the field will help regenerate it, improve its growth and regulate its harvest. Cultivating the tree will not only help to combat unsustainable exploitation, but will also meet the growing demand for treatments for a growing international health problem. *P. africana* was listed in Appendix II of the Convention on International Trade in Endangered Species of Wild Fauna and Flora (CITES) in 1995 at the request of Kenya. CITES export quotas for Cameroon have been in place since 2005 (Awono et al. 2016 and Mpouam et al. 2022). To fight against the overexploitation of this endangered species, its exploitation and marketing are now regulated in Cameroon (Tassiamba et al. 2022). This is not the case for many other endangered medicinal plant species. It is therefore important to train all those involved in the field of plant exploitation to be aware of the need for the conservation and sustainable exploitation of these plant species. YaBiNaPA organizes trainings for students, traditional healers, and heads of communities on the sustainable use of plants. Education and awareness about plant diversity, its role in sustainable livelihoods and importance to all life on earth are the top priorities of the hub. In some cases, the research hub encourages the cultivation of medicinal plants. YaBiNaPA is collaborating with the Ministry of Forestry and Wildlife to disseminate information on potentially active species and to propose measures for their conservation.

Cultural Exigencies

Cultural exigencies constitute one of the challenges the project has encountered. In many cultural communities in Cameroon, like in other parts of Africa, researchers are expected to comply with the cultural norms of the community before they are given the permission to carry out research. In order to obtain this cultural permission from the traditional ruler of the community, we often spend precious time to learn the cultural norms of the community, before taking an engagement to respect the tradition and customs of the people.

Most often, access to traditional knowledge has been a big challenge, as some indigenes do not disseminate the proper information needed, perhaps because of the fear that their knowledge, the basis of their culture, may be stolen. In spite of these cultural exigencies, YaBiNaPA researchers always carry out their investigations hitch-free, given their understanding that patience and respect of cultural norms are indispensable ingredients for the success of any research project.

Unnecessary Bureaucratic Hurdles from Customs and Management of Research Equipment

YaBiNaPA has also encountered some internal challenges involving research equipment. First, the process of getting the equipment out of the seaport or airport is always complex and unnecessarily long, despite the administrative support of the University of Yaounde 1 in the facilitation of the process. Second, the technical expertise and some machine parts required to repair the equipment is not always available in

Cameroon and we often solicit the assistance of international experts for the smooth functioning of the equipment, which may take a lot of time and money. Third, constant power failure is sometimes a major challenge for the smooth functioning of some of the equipment, especially those which cannot depend on the standby generator of YaBiNaPA as a source of energy.

Access to Chemicals and Reagents

Due to the quality of the chemicals used for analysis and research, most of the reagents are purchased abroad and it takes a lot of time for them to arrive. This aspect was well felt in YaBiNaPA during the COVID-19 pandemic where there was a general shutdown in the economy and only a few chemicals or reagents were received because some companies closed their offices present in Cameroon. This therefore slowed down research activities in YaBiNaPA. Fortunately, the situation is gradually coming back to normalcy.

5 YaBiNaPA and the Lessons Learned During the Collaborative Initiative

During this collaborative project, we have learned many lessons that are worth reporting in this paper. First, we learned that such a collaborative project cannot be successful in a sustainable manner if the State and administrative authorities are not active stakeholders. This implies that the success or failure of a collaborative project of this magnitude significantly depends on the political will and on the whims and caprices of policy makers, government officials, and other administrative stakeholders whose decisions can either "mar things" or "make things happen". This explains why, from the outset, YaBiNaPA has involved the State and other official authorities in every step of the project. The amazing success stories of YaBiNaPA today would not have been possible if the research hub did not have the full support of the government of Cameroon and that of the administrative authorities of the University of Yaounde 1 and the Ministry of Higher Education.

Second, we learned that the division of labour in a project with a huge societal impact, like YaBiNaPA, should include social scientists and the communities where the studies are conducted. In the context of YaBiNaPA, natural/medical and social scientists work in synergy, with well-defined roles, and the scientific outcome of such a close collaboration involving natural/medical and social scientists does not need to be over-emphasized. In addition, the involvement of community members in fieldwork activities was particularly beneficial and this significantly explains the success of YaBiNaPA scientific activities. In every case, YaBiNaPA always worked in collaboration with communities and policy makers to ensure the sharing of benefits, especially in the case of a breakthrough from its research activities. During field studies and meetings with communities, the results of the investigations of the species are presented. Thus, the species with a high potential of valorization are highlighted as well as those with high toxicities. Furthermore, the skills of members of the communities are also reinforced on the methods of sustainable harvest, culture, and the conservation of plants and derived products for the improvement of the life of the populations. The creation of collaborative platforms between YaBiNaPA, traditional healers, and community members has always facilitated access to scientific knowledge and data from plants research.

Third, we learned that there are many incredibly talented young scientists available, especially in Cameroon and Africa, who are not known and the reason is simply because they lack research opportunities and a conducive research environment for their amazing talents and ingenuity to be brought to the limelight. The junior scientists selected to join the research team of YaBiNaPA proved their ingenuity, hard work, assiduity, and professionalism beyond any reasonable doubts and significantly contributed to the success of all YaBiNaPA research and sustainable activities. The specific lesson we learned is that there are many great talents in Cameroon and Africa that have not been brought to the limelight, simply because they are not placed in a befitting research environment. The research facilities and opportunities offered by YaBiNaPA served as catalysts for the talents of these junior scientists to be unmasked. This implies that scientific ingenuity does not depend on race, region, or colour, but rather on the availability of research facilities and an enabling research environment.

Fourth, we learned that, if all the infrastructural resources and research equipment are available, teamworking and networking can produce amazing results, especially when everyone plays their role with dedication, assiduity, abnegation, and professionalism. This is the secret behind the different success stories of YaBiNaPA we are brandishing and celebrating today.

6 Conclusion: YaBiNaPA and the Way Forward

Cameroon, like many developing African countries, displays a precarious development landscape, characterized by different types of development-related problems, including abject poverty, numerous bacterial and parasitic diseases and the unaffordability of synthetic drugs, in spite of the fact that the country is a habitat for a very rich biodiversity of flora and fauna. Traditional healers (without any scientific knowledge) are exploiting these natural resources to provide healthcare to a large proportion of the population. The wish to explore and exploit the rich biodiversity of the country in a scientific manner to address the health challenges of Cameroon and those of other developing countries is one of the major motivations behind a South-North collaborative initiative between the University of Yaounde I (Cameroon) and Bielefeld University (Germany). The cooperation resulted in the conception of a research hub known as Yaoundé-Bielefeld Bilateral Graduate School of Natural Products with Antiparasite and Antibacterial Activity (YaBiNaPA), which is fully funded by the German Academic Exchange Service (DAAD), with funds from the German Federal Ministry for Economic Cooperation and Development.

The hub strives to involve traditional medicine (phytomedicine) into scientific investigation, provide scientific knowledge to traditional healers who rely on herbal plants to provide healthcare to a significant proportion of the population, and develop a phytodrug industry for economic, health, and scientific benefits and for the sustainable development of developing African countries in general and Cameroon in particular. In other words, this South-North collaborative initiative concentrates its scientific activities on traditional medicine to address the development challenges of developing African nations in general and those of Cameroon in particular, promote the United Nations Sustainable Development Goals (SDG), foster internationalization in higher education, and trigger the effective transfer of knowledge.

This paper has therefore assessed the activities of YaBiNaPA in the context of traditional medicine, by focusing on the goals of YaBiNaPA, its scientific achievements and knowledge transfer, its efforts to promote the United Nations SDG and international collaboration in higher education, its networking with health stakeholders, its outreach efforts, its implementation of the Nagoya Protocol in Cameroon, the challenges encountered, and the lessons learned from the collaborative initiative.

In light of the scientific endeavors of YaBiNaPA, the significant shift it has introduced to the traditional medicine sector in Cameroon, the positive influence it is having on the health industry, its strong support for the United Nations Sustainable Development Goals, the international partnerships it is cultivating in higher education, and its transformative effect on the development of Cameroon and Africa as a whole, it is essential for the hub to sustain its operations even after the conclusion of DAAD funding. This implies that the hub is in great need of new partners and new sponsors who can provide the funds necessary for it to remain in existence and continue carrying out its scientific activities for the sustainable and transformative development of Cameroon in particular and Africa in general.

References

Abia ALK, Ubomba-Jaswa E, Momba MNB (2015) Impact of seasonal variation on *Escherichia coli* concentrations in the riverbed sediments in the Apies River South Africa. Sci Total Environ 537:462–469

Akinyemi O, Oyewole SO, Jimoh KA (2018) Medicinal plants and sustainable human health: a review. Horticult Int J 2:194–195

Awono A, Eba'a Atyi R, Foundjem-Tita D, Levang P (2016) Vegetal non-timber forest products in Cameroon, contribution to the national economy. Int Forestry Rev 18:66–77

Balunas M, Kinghorn A (2005) Drug discovery from medicinal plants. Life Sci 78:431–441

Breda T (2014) Les délégués syndicaux sont-ils discriminés? Rev Économ 65:841–880

Butler MS (2004) The role of natural product chemistry in drug discovery. J Nat Prod 67:2141–2153

Calixto JB (2019) The role of natural products in modern drug discovery. An Acad Bras Ciênc 91(Suppl 3):e20190105

Ingram VJ, van Loo J, Vinceti B, Dawson I, Muchugi A, Duminil J, Sunderland T (2015) Ensuring the future of the pygeum tree (*Prunus africana*). Factsheet LEI Wageningen UR, 8p

Joyce JN (1993) The dopamine hypothesis of schizophrenia: limbic interactions with serotonin and norepinephrine. Psychopharmacology 112:S16–S34

Letouzey R (1985) Notice de la carte phytogéographique du Cameroun au 1:500000. 5 Vols. Institut de la carte Internationale de la Végétation, Toulouse

McHugh L (1996) World Health Organization: a fact sheet. https://digital.library.unt.edu/ark:/67531/metadc26072/

Mpouam AR, Bile AW, Wete E, Betti JL, Kouobegne S, Fouadjio BJ, Kourogue LR (2022) Etat des lieux de l'exploitation de *Prunus africana* (Hook. f.) *Kalkman* (Rosaceae) dans la région de l'Adamaoua, Cameroun: inventory of the exploitation of *Prunus africana* (Hook.f.) Kalkman (Rosaceae) in the Adamaoua region Cameroon. Int J Biol Chem Sci 16:1507–1521

Onana JM (2015) The world flora online 2020 project: will Cameroon come up to the expectation? Rodriguésia 66:961–972

Osunderu OA (2009). Sustainable production of traditional medicines in Africa. In: Yanful EK (ed) Appropriate technologies for environmental protection in the developing world. Springer, Dordrecht

Parasuraman S (2018) Herbal drug discovery: challenges and perspectives. Curr Pharmacogenomics Pers Med 16:63–68

Shu YZ (1998) Recent natural products-based drug development: a pharmaceutical industry perspective. J Nat Prod 61:1053–1071

Tassiamba SN, Betti JL, Temgoua LF, Kourogue LR, Bile AWN, Mpouam AR (2022) Evaluating the role of governance in the management of *Prunus africana* (Hook.f.) Kalkman in the Adamawa Region of Cameroon. Open J Forestry 12:408–431

Veldsema A, Van Gelder BF (1973) The ratio of type-1 and -2 Cu(II) in human ceruloplasmin. Biochim Biophys Acta 293:322–333

Norbert Sewald , born in 1961 in Munich (Germany), obtained his Ph.D. in Organic Chemistry at the Technical University of Munich. After a postdoctoral fellowship with Prof. J. E. Baldwin at the Dyson Perrins Laboratory, University of Oxford, he finished his habilitation in 1998 at the University of Leipzig and was appointed to Full Professor of Organic and Bioorganic Chemistry at Bielefeld University in 1999. His research interests comprise biocatalysis, the development of drug conjugates as well as the isolation, structure elucidation, and total synthesis of bioactive natural products (www.sewaldlab.de). He is the German co-cordinator of YaBiNaPA.

Marcel Frese studied Biochemistry at Bielefeld University. In 2011, he joined the Organic and Bioorganic Chemistry research group under supervision of Prof. Dr. Norbert Sewald, where he received his Ph.D. in 2015 on the exploitation of halogenating enzymes from natural product synthesis for biocatalytic purposes. Since then, he was project manager in several EU- and DAAD-funded research networks (www.magicbullet-reloaded.eu and www.yabinapa.de). His research interests focus on the application of enzymes for challenging reactions in organic synthesis.

Bruno Lenta Ndjakou , Professor at the University of Yaoundé I was born in 1973 in Bazou (Cameroon). He obtained his Ph.D. in Organic Chemistry at the University of Yaoundé I. His research interests include natural products chemistry, phytomedicine formulation, pharmacomodulation, drug discovery from plants and quality control of bioresources. He has been awarded a number of post-doctoral research grants such as Alexander von Humboldt Fellowship for Experienced Researchers (Bielefeld University, Germany), EU Marie Curie Internal Incoming Fellowship (Bielefeld University, Germany), Swiss National Science Foundation (University of Bern, Switzerland), Agence Universitaire de la Francophonie (University Louis Pasteur, Strasbourg, France). He received the Georg Forster Research Award by the Alexander von Humboldt Foundation (Germany) in 2022. He is the Cameroonian co-cordinator of YaBiNaPA.

Aloysius Ngefac , Full Professor of English Linguistics, is currently teaching at the University of Yaounde I, Cameroon. His research interests include sociolinguistics, World Englishes, and transformative development. He has received many prestigious awards such as Alexander von Humboldt Fellowship for Experienced Researchers (Germany), Fulbright Postdoctoral Scholarship (USA), and DAAD Fellowship (Germany) and given more than 15 different guest lectures in America and Germany, besides having published numerous scientific works and supervised five Ph.D. theses and more than 60 Master-level dissertations. He is a pioneer and current Book Series Editor for *Transformative Development for Postcolonial Africa* (Bloomsburg Publishing, Zed Books, London).

Cultivating Climate Resilience: Lessons from the CLIFOOD SDG Graduate School on Sustainable Agriculture, Interdisciplinary Education, and International Cooperation

Sintayehu Yigrem Mersha, Christian Brandt, Nicole Schönleber, Tesfaye Abebe Amdie, and Frank Rasche

Abstract Climate change poses serious challenges to food security, especially in countries like Ethiopia, where rain-fed agriculture and rapid population growth heighten vulnerability. Droughts, floods, pest outbreaks, and soil and water degradation are key threats to agricultural productivity. Addressing these issues, which have been reported as manifestations of climate change, requires international cooperation and capacity building, where higher education institutions play a vital role by generating knowledge, training leaders, and fostering collaboration. The Climate Change Effects on Food Security (CLIFOOD) SDG Graduate School, jointly coordinated by Hawassa University (Ethiopia) and the University of Hohenheim (Germany), supports the development of sustainable, climate-resilient agricultural practices. This chapter highlights lessons learned from CLIFOOD, emphasizing the importance of interdisciplinary education and international collaboration in tackling sustainability challenges. By uniting students and experts from diverse fields, CLIFOOD promotes integrated solutions that consider agricultural, environmental, and social dimensions. The program also showcases how internationalization can strengthen academic institutions. The chapter is organized around three questions: (1) What lessons emerged from establishing an SDG-aligned international project? (2) What are the key achievements and challenges in CLIFOOD's cooperation model? (3) How

S. Y. Mersha
School of Animal and Range Sciences, College of Agriculture, Hawassa University, Hawassa, Ethiopia
e-mail: sintayehu@hu.edu.et

C. Brandt (✉) · N. Schönleber
Institute of Farm Management, University of Hohenheim, Stuttgart, Germany
e-mail: christian.brandt@uni-hohenheim.de

T. A. Amdie
School of Plant and Horticultural Sciences, College of Agriculture, Hawassa University, Hawassa, Ethiopia

F. Rasche
International Institute of Tropical Agriculture (IITA), Nairobi, Kenya

C. Brandt et al. (eds.), *Implementing Sustainability and Internationalization in Higher Education*, World Sustainability Series, https://doi.org/10.1007/978-3-031-97837-1_8

do bilateral graduate schools like CLIFOOD advance sustainability and support the Sustainable Development Goals (SDGs)?

1 Introduction

The global community is currently grappling with the multifaceted challenges of climate change—challenges that have profound implications for food security. Rising global temperatures, erratic rainfall patterns, and an increasing frequency of extreme weather events such as droughts, floods, and pest outbreaks (e.g., desert locusts) are severely impeding the productivity and profitability of agricultural systems. These phenomena are especially damaging in regions that rely largely on rain-fed agriculture, such as Eastern Africa. In countries like Ethiopia, where smallholder farmers dominate, shifting growing seasons, water scarcity, and soil degradation have led to significant losses in both crop production and livestock farming. As a result, millions of people are at heightened risk of hunger and malnutrition, emphasizing the urgent need for effective and sustainable agricultural solutions.

In response to these pressing challenges, the CLIFOOD (Climate Change Effects on Food Security) SDG Graduate School was launched in 2016 under the framework of the Sustainable Development Goals (SDG) Graduate Schools. This initiative was born from a strategic collaboration between Hawassa University in Ethiopia and the University of Hohenheim in Germany, exemplifying how international academic cooperation can drive transformative change. CLIFOOD's mission is to build capacity and drive research that addresses food insecurity in the context of climate change, while enhancing the sustainability and resilience of agricultural systems. By integrating teaching, research, and institutional strengthening, the program seeks to deliver long-term impact at both local and global levels. The CLIFOOD Graduate School is part of the broader DAAD SDG Graduate School Program, funded by the German Federal Ministry for Economic Cooperation and Development (BMZ). This program strengthens educational and research capacity in the Global South by training specialists at the master's, Ph.D., and postdoctoral levels through partnerships with universities in Africa, Latin America, and Asia.

Higher education institutions are uniquely positioned to contribute to sustainable development by generating critical knowledge, training future leaders, and fostering cross-border collaborations. In the case of CLIFOOD, its focus on climate-resilient food systems and alignment with the Sustainable Development Goals enable universities to address the intertwined challenges of climate change, food insecurity, and poverty. Here, sustainability extends beyond environmental protection to encompass economic and social dimensions, ultimately empowering communities and shaping a more resilient future.

A central element of CLIFOOD is its interdisciplinary approach. By promoting collaboration among students, researchers, and experts from diverse fields,

CLIFOOD ensures that the solutions developed consider a broad range of perspectives, spanning from agricultural and environmental to social and economic dimensions. This holistic approach is critical for addressing the complex challenges posed by climate change, especially in regions that are disproportionately affected, such as Ethiopia. Furthermore, the internationalization aspect of CLIFOOD has played a key role in strengthening academic institutions. The South-North collaboration has allowed Ethiopian researchers to gain access to advanced methodologies and technologies, while German researchers have benefited from direct insights into real-world challenges faced by communities in East Africa. Such bilateral exchanges not only enhance research capabilities but also foster a deeper mutual understanding that is essential for creating solutions that are both globally informed and locally relevant.

Joint research initiatives and capacity-building workshops have further solidified the partnership between the two institutions. These efforts have enhanced the technical and analytical skills of participants, ensuring that local institutions are well-prepared to address climate-related food security challenges long after the project's conclusion. The development of human capital and the pooling of resources serve as a testament to how international cooperation can drive sustainable change.

This book chapter draws on the experiences of CLIFOOD and is structured around three pivotal questions: 1. Key Lessons: What are the primary lessons learned from establishing an international project like the SDG Graduate Schools, which align with a global agenda such as the SDGs? 2. Achievements and Challenges: What are the principal achievements and challenges encountered in implementing such international cooperation? 3. Impact on Sustainability: How do bilateral graduate schools contribute to advancing sustainability and achieving the SDGs?

This chapter begins by examining the global impacts of climate change on agriculture and food production, with a focus on Ethiopia. It then details how institutional arrangements, and international collaborations have been leveraged to respond to these crises. Using the genesis and evolution of the SDG Graduate School—and CLIFOOD in particular—as a case study, the discussion explores the key pillars that have underpinned the initiative, such as interdisciplinary education, capacity building, and knowledge exchange.

Subsequent sections provide a detailed account of the lessons learned from CLIFOOD, highlighting both successes and the obstacles encountered along the way. The narrative also underscores the importance of internationalization in enriching academic programs and research endeavors, ultimately contributing to a more resilient and equitable future. Through this comprehensive examination, this chapter illustrates how sustainability and internationalization are not only compatible but mutually reinforcing pillars in modern higher education.

2 Background

Climate Change, Food Security, and Agriculture

Human activities, particularly the emission of greenhouse gases, have unequivocally driven global warming. Between 2011 and 2020, the global surface temperature increased by 1.1 °C above pre-industrial levels (circa 1850–1900) (IPCC 2023). This rise is primarily attributed to escalating greenhouse gas emissions, which are in turn linked to unsustainable energy practices, inefficient land use, and divergent consumption and production patterns at both global and national levels. Climate change affects a broad range of systems that underpin human well-being—including water resources, agriculture, forestry, coastal and marine ecosystems, human settlements, and public health (IPCC 2023). These sectors encompass both natural and human-managed components, highlighting how interconnected our environment and societal systems are.

The agricultural sector is particularly sensitive to these changes. Even slight variations in temperature, precipitation, or atmospheric CO_2 concentrations can significantly alter plant growth, thereby influencing the cultivation of food, feed, and fodder. Factors such as excessive or insufficient rainfall, extreme temperatures, changes in the length of the growing season, and other severe weather conditions all contribute to fluctuations in food availability, market prices, and overall food security.

In many low-income countries, agriculture constitutes at least 40% of the gross domestic product (GDP) and employs about 80% of the labor force. Additionally, around 70% of the world's impoverished population, largely dependent on agriculture, reside in rural areas (Byerlee et al. 2008). A significant portion of this rural population lives in arid or semi-arid regions, which are especially vulnerable to climate variability. Extreme events, such as droughts and heat waves—which are increasing in frequency and intensity due to climate change (IPCC 2023)—further exacerbate the vulnerability of these rural livelihoods. This increased susceptibility not only undermines the availability of food but also affects its access, utilization, and stability—the four pillars of food security. By altering environmental conditions and weather patterns, climate change disrupts these pillars, creating a complex challenge to global food security (FAO 2018).

Ethiopia's Struggle with Climate Change and Food Security

Agriculture is the cornerstone of Ethiopia's economy, contributing 33% of GDP (National Bank of Ethiopia 2020) and providing livelihoods for about 67% of the labor force. Most rural communities depend on rain-fed, small-scale agriculture to produce food, fiber, and energy (World Bank 2020a, b, c), making the sector critical for both food security and employment. However, climate change poses a significant threat to this sector. Projections indicate that climate change could reduce GDP growth by 0.5–2.5% per year (World Bank 2010) and may lead to an overall GDP decline of around 10% by 2045, primarily due to decreased agricultural productivity (USAID 2016). Ethiopia cultivates a total of 14.5 million hectares for crop production. Of this, cereal crops (Teff (20% of total area under cereal crops), Maize (17%),

Wheat (13%), Sorghum (12%), and Barley (6%)) dominate with approximately 10 million hectares, representing 69% of the total cultivated land (CSA 2020). Pulses, on the other hand, are grown on about 1.67 million hectares (CSA 2020). In this category, faba bean (0.52 million hectares) and haricot (common) bean (0.34 million hectares) are dominating. Additionally, coffee—the country's main cash crop—is grown on 0.86 million hectares, accounting for about 6% of the cultivated area.

Overall, the estimated total production of crops from the land cultivated in 2021/22 was 29.1 million tons of cereal grains and 3.17 million tons of pulses. Enset, a perennial herbaceous crop and staple food for over 20 million people in Southern Ethiopia, is grown on a relatively small area—about 0.4 million hectares (approximately 3% of the national cultivated area). Unlike most annual crops, enset production is typically quantified in terms of its two main products—Kocho, a fermented starch-rich food made from the decorticated and pounded pseudostem and corm, and Bulla, a fine white starchy powder extracted from the squeezed pulp—rather than solely by cultivated area. Detailed CSA survey data (2020) indicate that enset is cultivated on roughly 157.7 million plants, which yield around 48.9 million quintals of Kocho (equivalent to about 4.89 million metric tons) and approximately 1.48 million quintals of Bulla (roughly 148,000 metric tons). These figures highlight enset's remarkable productivity per unit area, underlining its capacity to feed a large population despite its limited cultivation area. In this context, it has to be bearded in mind that some information sources report enset dry matter production as high as 13.4 million tons, which may reflect differences in measurement and estimation methods.

Agriculture in Ethiopia faces numerous challenges that contribute to a significant yield gap—the difference between actual production and the potential yields achievable under optimal conditions, such as sufficient rainfall, fertile soil, quality seeds, and proper farming practices. Factors such as soil fertility depletion, water and wind erosion, drought, inappropriate crop and variety selection, and climate change all play significant roles. While production and productivity of some crops have increased over the past decade, these improvements have not kept pace with the demands of a growing population. As a result, Ethiopia remains a net importer of agricultural commodities—including wheat, palm oil, cane sugar, rice, and sunflower seed oil—even though most of these could be produced locally to conserve foreign exchange. Although the share of imported food varies by commodity, Ethiopia remains a net importer of agricultural goods despite its enormous potential. For example, Ethiopia imports roughly one million metric tons of wheat every year, underscoring its reliance on imports to bridge the gap between domestic production and consumption needs (U.S. Department of Commerce 2023). This significant dependency highlights the strength of the dilemma and underscores Ethiopia's urgent need to restore its self-sufficiency.

There is considerable potential to enhance crop yields by developing and implementing context-specific best agricultural practices tailored to the diverse agroecological zones of Ethiopia. Projects such as CASCAPE (Capacity Building for Scaling up of Evidence-based Best Practices in Agricultural Production of Ethiopia) and REALISE (Realising Sustainable Agricultural Livelihood Security in Ethiopia) are long-term, comprehensive capacity-building initiatives. They focus on documenting, disseminating, and scaling up best agronomic practices—including the use of optimal crop varieties and appropriate fertilizer applications—in alignment with local agroecological and socioeconomic conditions. These projects, executed by partnerships between Ethiopian research institutions (such as Jimma University, Haramaya University, Bahir Dar University, Hawassa University, Mekelle University, and Addis Ababa University) and Dutch experts from Wageningen University and Research Centre, are financially supported by the Government of the Kingdom of the Netherlands through its Embassy in Addis Ababa, under the broader BENEFIT partnership portfolio. REALISE, for instance, builds on the achievements of CASCAPE and other programmes like ISSD Ethiopia to extend their impact to new areas and target groups, demonstrating that integrated and sustained capacity-building efforts can significantly improve current low agricultural yields—potentially even doubling them when best practices are adopted. Research outputs from CLIFOOD will considerably contribute to the development of climate-resilient and sustainable agricultural practices, ultimately enhancing food security in the region.

Livestock are integral to Ethiopia's farming systems. Smallholder farmers depend on livestock not only for plowing with oxen but also for the production of meat, milk, and eggs, as well as for additional income. Ethiopia leads Africa in livestock numbers, with 70.3 million cattle, 52.5 million goats, 42.9 million sheep, and 8.9 million camels (CSA 2021). Despite these impressive livestock numbers, productivity remains suboptimal—production levels for milk, meat, and eggs fall significantly short of the sector's full potential. Annual production is reported at 4.3 billion litres of milk, 294 thousand tonnes of meat from cattle, sheep, goats, and camels, 48 thousand tonnes of chicken meat, and only 2. billion eggs—figures that fall short of the sector's potential (Legese et al., 2023).

Climate variability and change exacerbate these challenges, threatening a large proportion of the rural population (Araro et al. 2020). Rural farmers, often with limited access to education, technology, and essential support services, are constrained in their ability to adapt to these changes. Natural phenomena such as El Niño contribute to erratic rainfall and droughts, while unprecedented seasonal irregularities disrupt agricultural cycles. For instance, the 2015–2016 El Niño event led to delayed and erratic kiremt rains, resulting in acute and widespread crop failure, asset depletion, and food insecurity across large parts of the country (OPM and HESPI 2018). Similarly, the 2015 kiremt season experienced below-average rainfall, affecting north and central Ethiopia and leading to one of the worst droughts in decades. These disruptions have had severe impacts on agriculture and livelihoods, particularly for rural farmers who rely heavily on predictable rainfall patterns (Philip et al. 2018). This situation is further aggravated by climate-induced biodiversity loss, which weakens ecosystem resilience and diminishes the productivity of natural services—such as pollination and natural pest control. This loss not only undermines food security but also complicates the implementation of sustainable agricultural practices in Ethiopia. For instance, the recent desert locust outbreak in East Africa (February 2020) severely impacted food production, with assessments by the Ethiopian Government and FAO (World Bank 2020c) estimating that over one million people in Ethiopia required food assistance. The outbreak resulted in an expected loss of about 356,000 tonnes of cereals—including sorghum, maize, and wheat—along with damage to approximately 197,000 ha of arable land and 1,350,000 ha of pasture (WHO 2022).

Recurrent droughts and delayed or inadequate rainfall have forced communities into reactive and unsustainable coping mechanisms that lack long-term resilience. Practices such as overgrazing, cultivating unsuitable lands, increased charcoal production, and deforestation not only fail to address immediate needs but also contribute to ongoing environmental degradation, further diminishing the region's capacity to withstand future extreme events (Matewos 2019).

Addressing these multifaceted challenges requires proactive measures. Adapted planning and early, reliable weather forecasting are crucial to anticipate abnormal rainfall patterns and other climatic events. The selection of climate-resilient crop and livestock varieties, along with improved agricultural inputs and practices, will be vital for enhancing productivity. Additionally, strategies that focus on soil and biodiversity conservation, adaptive livestock systems, and holistic approaches to food security are essential. Evidence-based recommendations must support Ethiopia's rapidly growing population—now estimated at around 130 million, making it the second most populous country in sub-Saharan Africa. Special attention is required for vulnerable groups, particularly women and children. In Ethiopia, women often face diminished decision-making power and reduced economic security, while children are especially vulnerable to the physiological stresses of extreme weather events and climate-related disasters. Consequently, both groups are at increased risk during such emergencies (Pritchard et al. 2016; Akachi et al. 2009; UNICEF 2007).

Through a combination of improved agricultural practices, targeted research and education initiatives like CLIFOOD, and proactive policy measures, there is an opportunity to transform Ethiopia's agricultural sector into one that is both resilient and sustainable in the face of a changing climate.

Tackling Climate Change in Ethiopia: General Pathways

In 2016, the Ethiopian government committed to contribute to the UN Sustainable Development Goals (SDGs 2015) by 2030 (Ethiopia 2017 Voluntary National Review on SDGs, 2017). This integration was notably reflected in the Second Growth and Transformation Plan (GTP II), which aligns with the objectives of the 2030 Agenda for Sustainable Development. The GTP II serves as a strategic plan to guide Ethiopia's development efforts in harmony with the SDGs. An official report reviewed the country's efforts to address six key SDGs—No Poverty (Goal 1), Zero Hunger (Goal 2), Good Health and Well-being (Goal 3), Gender Equality (Goal 5), Industry, Innovation and Infrastructure (Goal 9), and Life Below Water (Goal 14)—which have been systematically integrated into national strategies such as the Growth and Transformation Program (GTP II). As a result, Ethiopia's national poverty rate declined from 44% in 1999/00 to 24% in 2014/16, with projections indicating a further reduction to 17% by the end of GTP II in 2019/20 (National Planning Commission 2017).

Despite these notable gains, the 2024 Global Multidimensional Poverty Index (OPHI and UNDP 2024) reveals that Ethiopia is still home to approximately 86 million people (70% of the population) living in multidimensional poverty—a measure that captures multiple deprivations in health, education, and living standards beyond simple income levels. This places Ethiopia among the five countries with the highest absolute numbers of poor people globally, alongside India (234 million, 17%), Pakistan (93 million, 39%), Nigeria (74 million, 34%), and the Democratic Republic of the Congo (66 million, 70%). The high poverty incidence is driven by persistent deprivations in essential areas such as housing, sanitation, education, and nutrition, as well as low Human Development Index scores. Rural areas, where most Ethiopians reside, are particularly affected, with significantly higher poverty rates than urban areas.

These trends underscore the urgent need for a comprehensive strategy that goes beyond merely addressing climate change impacts. Empowering local experts to drive sustainable solutions is essential. This effort involves training a new generation of scientists, expanding research and teaching infrastructure, and developing innovative curricula designed to foster practical solutions for food security and sustainable agriculture. The ultimate objective is to build a resilient, knowledge-based agricultural sector capable of adapting to climate variability, implementing transformative practices, and securing the livelihoods of communities facing these multifaceted challenges.

3 Institutional Collaborations and Responses

History and Expertise: UHOH and HU in Agriculture

UHOH was established in 1818 in the aftermath of the catastrophic famine triggered by the 1815 eruption of the Tambora volcano in Indonesia. The eruption caused significant climatic disturbances, culminating in the "Year Without a Summer" in 1816, during which Europe experienced widespread crop failures and livestock fatalities, leading to one of the biggest famines in Baden-Würtemberg. In response, King Wilhelm I and Queen Katharina Pavlova founded the Hohenheim Agricultural Teaching and Experimental Institute in Baden-Württemberg, with the primary mission of enhancing agricultural productivity and ensuring food security. Over the past two centuries, UHOH has earned a formidable reputation in agricultural research, particularly in food security and climate change. In recent years, UHOH has broadened its international research and education scope, aligning its programs with global agendas such as the UN SDGs. It offers a range of training programs at both undergraduate levels—such as Agricultural Sciences, Earth System Science, and Environmental Protection and Agricultural Food Production—and postgraduate levels, including Global Food Security, Water for Life, and Tropical Agriculture, Food and Resource Sciences.

HU, one of Ethiopia's 51 public universities and recognized as one of its eight research universities, is home to the College of Agriculture, which was established in 1976. As the oldest and a founding institution within HU, the College of Agriculture plays a central role in the university's academic structure. HU offers more than 300 specialization programs across Ph.D., MSc/MA, and BSc/BA levels. The College of Agriculture comprises four schools that deliver diverse programs in crop and horticulture production, animal husbandry, gender studies, environment and development studies, human nutrition, and food science and technology. With an enrollment ranging between 28,000 and 43,000 students across various programs, HU boasts a strong research culture in agricultural sciences, having developed several improved crop varieties and animal breeds vital for national food security. Additionally, HU offers specialized academic programs on climate change and sustainable agriculture through centers of excellence such as the Centre of Excellence for Human Nutrition, the Centre for Policy and Development Studies, and the Center for Ethiopian Rift Valley Studies. Currently, HU collaborates on over 81 projects with universities across Africa, Europe, Canada, and the United States, with CLIFOOD being a prominent example. Agriculture remains a primary focus at HU, seamlessly integrating education, research, community service, and strategic partnerships.

The Evolution of UHOH-HU Collaboration and CLIFOOD's Future

The official cooperation between UHOH and HU has a long history, predating the establishment of CLIFOOD. For years, several institutes at UHOH—most notably the Hans-Ruthenberg Institute of Agricultural Sciences in the Tropics and the former CLIFOOD host, the DAAD Centre of Excellence, now known as the Food Security Centre (FSC)—have collaborated with various HU institutes. This collaboration primarily involved the office of the Vice President for Research and Technology Transfer and multiple departments within the College of Agriculture.

Despite significant expansion and achievements since its establishment as a comprehensive university in 2000, HU continues to face challenges in producing highly qualified professionals and researchers in the agricultural field. There remains a critical need for ongoing training and capacity building for its staff, particularly since some programs require professionals with advanced qualifications, such as a Ph.D. Key issues include enhancing quality management, improving teaching standards, and optimizing education and research through multidisciplinary approaches. Ultimately, by cultivating highly qualified experts for future labor markets in alignment with the UN Agenda 2030's commitment to "leave no one behind," CLIFOOD represents a strategic initiative to address critical gaps in capacity and quality, thereby strengthening both educational and research outputs in the agricultural sector. From HU's perspective, CLIFOOD was established with the following expectations:

1. Alignment with Global Agendas: Align study programs with global initiatives such as the UN SDGs.
2. Enhanced Knowledge Sharing: Increase the exchange of expertise to boost research and educational outcomes.
3. Innovative Teaching and Capacity Building: Promote innovative teaching methods, including eLearning, along with various capacity-building approaches.
4. Development of Sustainable Technologies: Foster the creation of sustainable agricultural technologies and practices that improve food security.

5. Strengthened Internationalization: Enhance international visibility and further cooperation between institutions.

4 CLIFOOD—Addressing Food Security and Climate Change

The establishment of CLIFOOD primarily addresses two critical regional agendas in Eastern Africa: food security and climate change. Central to addressing these challenges is imparting fundamental knowledge about the causes and consequences of climate change—a core component of CLIFOOD's integrated research and teaching programs. While direct beneficiaries are students and researchers, the broader rural population, who are most vulnerable to climate impacts, also stand to gain.

Aligned with the UN Agenda 2030's commitment to 'leave no one behind' and its focus on 'quality education' and 'zero hunger', the CLIFOOD are designed to generate and disseminate practical, solution-oriented knowledge for policymakers, civil society, and other key stakeholders. CLIFOOD's work addresses several SDGs, with a particular emphasis on eradicating hunger and poverty, safeguarding human dignity, promoting equality and health, and ensuring environmental protection.

CLIFOOD is built on three key pillars—*Teaching, Research, and Capacity Building*—each playing a critical role in advancing education, fostering innovative research, and strengthening resilience to address food insecurity and climate change in line with the SDGs. CLIFOOD focuses on training a new generation of scientists and leaders through a comprehensive, interdisciplinary approach.

Teaching

CLIFOOD offers an extensive educational framework that integrates interdisciplinary curricula spanning agricultural, natural, environmental, nutritional, and food sciences. Currently in its second phase (2021–2025), CLIFOOD has awarded 13 Ph.D. and 2 postdoctoral long-term scholarships, enabling students to conduct SDG-related research focused on food security and climate change. In the first phase, a total of 12 Ph.D. and 2 postdoctoral scholarships were granted, making the overall count 25 Ph.D. and 4 postdoctoral scholarships (see Table 1).

Teaching modules are co-developed by Ethiopian and German scientists and are designed to provide high-quality education. The interdisciplinary nature of these modules ensures that students gain a deep understanding of the multifaceted challenges at the intersection of climate change and food security. In addition to the core scientific curriculum, students participate in up to six digital transferable skills courses, which enhance both their academic and practical competencies. CLIFOOD also features block seminars, workshops, and interactive teaching sessions—often with international experts—to help students place their learning within a global sustainable development framework.

Table 1 Overview of research projects in the CLIFOOD SDG graduate school

Research project titles
Robust adaptation strategies to climate change for Ethiopian agriculture—heterogeneity of farming households and the role of social networks
Grain quality characteristics of Ethiopian barley genotypes as affected by climate change
Exploitation of cassava for food and feed
Control of the weed *Striga hermonthica* by the fungal biocontrol agent *Fusarium oxysporum* f.sp. *strigae*
Reducing uncertainty in predictions of climate change impacts on agricultural crop production in Ethiopia
Seasonal forecast of weather extremes for improving food security
Climate change and intergenerational persistence of poverty and malnutrition
Use of plant growth models and seasonal weather forecast for yield stabilization under vulnerable crop cultivation conditions
Grain yield and quality responses of wheat genotypes under CO_2 enrichment and extreme events
Advanced seasonal forecasts for the Horn of Africa by optimized initialization and perturbation with satellite data
Exploitation of cassava for food and feed—Implementation
Ecological interactions of cereals with root-associated microorganisms with crop health promoting feature
Agricultural intensification for improving soil health, enhancing food security and climate change adaptation in Southern Ethiopia
Seed regeneration of invasive species: germination ecophysiology and woody vegetation encroachment processes in Borana, southern Ethiopia
Effect of dietary interventions in climate sensitive agriculture target villages in Southern Ethiopia on micronutrient status of mothers and children
Assessing the feeding values of potential local feed resources with low enteric methane emission in vivo for improved livestock productivity and sustainable food security in Ethiopia
Challenges, opportunities and adaptation strategies of sorghum production under climate change in southern Ethiopia
Grain legumes for climate change and food security
Agrobiodiversity and seed systems of indigenous root and tuber crops in Ethiopia: biological and social adaptive capacity of smallholder farming systems
Production, reproduction and adaptation traits of the Boran cattle breeds under changing climate in the lowlands of Ethiopia
Effect of climate change and weather variability on child nutrition and health in Ethiopia
Possibilities for raising wheat (*Triticum aestivum* L.) productivity to improve food security under climate change
Mining natural variation for improvement of *Plectranthus edulis*/yam: analysis of genetic diversity and selection on phenotypes and genes for yield, stress resistance and ecological adaptation

(continued)

Table 1 (continued)

Research project titles
Isolation and characterization of N-fixing and phosphate solubilizing bacteria from grain legumes: for sustainable nutrient management and crop productivity during the era of climate change
Study on soil fertility, nutrient balances and biological N-fixation under contrasting soil pH and cropping systems in southern Ethiopia
Validating the potentials of unconventional feeds on meat and egg production and qualities of poultry for enhanced food security through supply of high value protein in Ethiopia
Evaluation of the feeding value, tannin deactivation, anti-parasitic effects and methane emission reduction potential of selected browse and/or herbaceous legume species fed to sheep/goats in Ethiopia
Inter- and transdisciplinary data center to improve agricultural practice and management, and to predict extreme events
Best practices and perspectives for scaling up agricultural innovations based on the CLIFOOD research results

Research

CLIFOOD's research agenda directly addresses the need to mitigate the impacts of climate change on food systems, with a primary focus on Ethiopia and Eastern Africa. Since it's beginning all research positions are exploring critical issues in the fields of crop science, livestock systems, meteorology, and soil management. These efforts aim to develop resilient agricultural strategies capable of withstanding climate-related disruptions.

A key strength of CLIFOOD's research strategy is its demand-driven approach, ensuring that scientific inquiries are closely aligned with the region's most pressing development challenges. Emphasizing climate-smart agricultural practices, CLIFOOD contributes directly to the objectives of the UN Agenda 2030. Scholars benefit from South-North and South-South academic exchanges, enabling collaboration with scientists worldwide and the application of innovative methods in local contexts. Through close collaboration between Ethiopian and German institutions, the research agenda fosters interdisciplinary and transdisciplinary projects that not only deepen scientific understanding but also offer practical solutions to the intertwined challenges of food insecurity and climate resilience.

Capacity Building

Capacity building is a cornerstone of CLIFOOD's mission, extending its impact beyond academia to benefit both academic and administrative staff. CLIFOOD offers interdisciplinary training that supports professional development in research, teaching, and administration, preparing participants for leadership roles in academia, industry, and policymaking. Findings from tracer studies of Phase 1 alumni highlight notable career advancements, for example: 1. Several CLIFOOD graduates have secured postdoctoral positions at internationally renowned agricultural research centers in Addis Ababa, where they contribute to cutting-edge research in crop improvement., 2. A number of CLIFOOD alumni have transitioned into academic roles, holding assistant or associate professorships at leading Ethiopian universities where they are actively involved in teaching, research, and administration., 3. Other CLIFOOD graduates are now working as agricultural scientists within national research institutions, where they help develop innovative practices to enhance food security and sustainable agriculture.

These outcomes underscore the program's impact in building a cadre of highly qualified professionals capable of driving sustainable development in the region. Block seminars and soft skills training—including courses in conflict management, scientific writing, project management, and intercultural communication—play a vital role in these efforts.

ICT-based teaching methods are employed to create a modern, flexible learning environment that is responsive to the evolving educational needs of scholars, particularly in developing countries. CLIFOOD's emphasis on multidisciplinary collaboration unites experts from agronomy, animal husbandry, soil science, agricultural economics, biotechnology, food science, and other fields. This collaborative framework addresses the complex issues of food security and climate change by enabling scholars to apply insights from various disciplines to real-world problems. Furthermore, joint supervision by Ethiopian and German scientists ensures that research projects remain locally relevant while also contributing to global scientific discussions.

5 Aligning the CLIFOOD SDG GS Program with Global Agenda

CLIFOOD has made significant contributions to achieving the SDGs through its interdisciplinary approach, innovative research, and capacity-building efforts. Some of the key lessons learned from the program's contributions to specific SDGs are outlined below:

Addressing Poverty through Climate-Smart Agriculture (SDG 1—No Poverty)

CLIFOOD demonstrates that building resilience to climate change is crucial in reducing poverty among smallholder farmers. By employing seasonal high-resolution forecasts and bio-economic analyses, CLIFOOD shows that climate-resilient agricultural practices can improve productivity and mitigate climate-related risks, ultimately helping vulnerable communities avoid poverty (Ayalew et al. 2022a, b, c; Bayssa et al. 2021; Rettie et al. 2022; Senbeta et al. 2024; Yimer et al. 2022).

Combatting Hunger with Sustainable Agricultural Practices (SDG 2—Zero Hunger)

CLIFOOD emphasizes the importance of utilizing underexplored crops, improving livestock systems, and enhancing agricultural sustainability to address food insecurity. Through research on drought-tolerant crops and protein extraction from local resources like cassava leaves, CLIFOOD highlights the value of context-specific agricultural innovations that sustainably increase food production and secure nutrition (Ayalew et al. 2021; Bayssa et al. 2021; Belete et al. 2024; Dejene et al. 2022; Yimer et al. 2022).

Enhancing Health through Improved Nutrition (SDG 3—Good Health and Well-being)

CLIFOOD's research establishes a direct link between food security and improved health outcomes. By promoting diversified and resilient diets and reducing child malnutrition, CLIFOOD underscores the critical role that agricultural research plays in enhancing public health in areas vulnerable to climate-induced food insecurity (Ayele et al. 2022; Desta et al. 2024a, b).

Promoting Sustainable Development through Education (SDG 4—Quality Education)

CLIFOOD integrates advanced, interdisciplinary education programs that address complex global challenges such as food security and climate change. Through block seminars, transferable skills courses, and a holistic approach to farming systems, CLIFOOD develops a new generation of scientists equipped to tackle these issues, reinforcing the pivotal role of quality education in sustainable development.

Advancing Gender Equality in Agricultural Research (SDG 5—Gender Equality)

CLIFOOD has been explicitly committed to promoting equal opportunities throughout its implementation. In the call for applications for the second funding phase, women were strongly encouraged to apply. Recognizing the underrepresentation of women in the first phase—where only 3 out of 14 scholars were female (approx. 21%)—CLIFOOD management advised subproject leaders to pay special attention to gender balance when selecting candidates, assuming equal qualifications. This led to a slight increase in female participation in the second phase, with 4 out of 16 scholars being women. In total, 7 out of 30 Ph.D. and postdoctoral scholars supported by CLIFOOD were female. In 2023, CLIFOOD's targeted efforts yielded visible results: a postdoctoral position was awarded to a highly qualified female scientist, and three women applied for a vacant Ph.D. position. These outcomes reflect a growing awareness of the need for more inclusive academic environments.

Sustainable Resource Management to Support Clean Water and Sanitation (SDG 6—Clean Water and Sanitation)

CLIFOOD's work in soil conservation, erosion control, and sustainable water use highlights the interconnectedness of agriculture, water resource management, and environmental sustainability. By promoting practices that conserve water and improve soil fertility, CLIFOOD plays a crucial role in protecting water resources and supporting clean water initiatives (Habte et al. 2021; Ware et al. 2022; Yimer et al. 2024a, b; Dejene et al. 2022).

Fostering Climate Action Through Knowledge Transfer and Innovation (SDG 13—Climate Action)

CLIFOOD builds local and regional capacities for climate action by combining scientific research, data-driven approaches, and practical implementation. The establishment of a scientific data center and the development of innovative agricultural practices exemplify how CLIFOOD's cross-disciplinary research can significantly enhance the region's ability to adapt to climate change (Habte et al. 2020; Gardi et al. 2022a; Gardi et al. 2022b; Kayamo et al. 2023; Kebede et al. 2024; Rettie et al. 2023b).

Conserving Terrestrial Ecosystems with Sustainable Farming Practices (SDG 15—Life on Land)

CLIFOOD promotes agro-ecological practices—such as intercropping and the use of legumes—that improve soil health and enhance pest and disease control. By demonstrating that sustainable farming techniques can boost productivity while protecting ecosystems, CLIFOOD underscores the dual benefits of environmental conservation and enhanced agricultural resilience (Tadesse et al. 2022; Yimer et al. 2022; Abate et al. 2024a, b).

Strengthening Global Partnerships for Sustainable Development (SDG 17—Partnership for the Goals)

CLIFOOD's international collaboration between Ethiopian and German institutions exemplifies the power of global partnerships in addressing complex challenges. Through networking among SDG graduate schools and collaborations with local and international stakeholders, CLIFOOD enhances knowledge exchange and capacity building, demonstrating the critical role of multi-stakeholder cooperation in achieving sustainable development. In 2024 alone, CLIFOOD worked with over 17 institutional partners across academia, government, civil society, and international organizations. It engaged 23 external experts in seminars and skills courses, hosted stakeholder events with over 100 participants—including farmers, researchers, and policymakers—and contributed to international conferences, reinforcing its commitment to meaningful knowledge exchange.

CLIFOOD's strategy and experience illustrate that integrated, interdisciplinary research and capacity-building efforts are key to addressing SDG-related challenges. By promoting climate-resilient agricultural practices, advancing gender equality, fostering innovation through education, and strengthening global partnerships, CLIFOOD demonstrates that comprehensive, locally relevant solutions can make significant strides toward achieving the SDGs.

6 Successes in Running Interdisciplinary Programs

In CLIFOOD, Ph.D. scholars enrolled in a series of specialized courses that complemented the regular curriculum of their respective faculties. In addition to semester-long courses, scholars participated in structured block seminars and soft skill training sessions—unique offerings of CLIFOOD's education framework.

The curriculum included interdisciplinary block seminars, each spanning a week and alternatingly hosted in Ethiopia and Germany. These seminars, designed to encourage scholars to explore topics beyond their specialized fields, have featured lectures from over 100 experts drawn from international and national organizations across multiple countries. This diverse teaching structure equips graduate students to think holistically, understanding the interconnections among agriculture, food security, climate change, and sustainable development while aligning with the global sustainable development agenda. In response to challenges such as the COVID-19 pandemic and security issues in Ethiopia, CLIFOOD integrated ICT methods, including blended learning and live video sessions, to maintain educational quality and continuity.

The thematic focus of the block seminars covered a broad range of disciplines, including food and nutrition security, climate modeling, agronomy, livestock intensification, socio-economic development drivers, and natural resource degradation. These topics were formulated by a designated scientific leader involved in CLIFOOD, and experts—sourced from both within and outside the CLIFOOD network—were invited to present their insights for a given topic. Each block seminar featured presentations by up to 10 experts, incorporating keynote addresses and specialized professional sessions. Scholars, regardless of their specialization, were required to develop a comprehensive understanding of these interdisciplinary topics. Attendance at these block seminars was mandatory for receiving the graduate school's certificate, which is awarded in conjunction with the Ph.D. degree.

CLIFOOD also leveraged its Food Security Centre (FSC) network, drawing on experts from various disciplines for both qualification and soft skill development programs. Experts from countries such as Germany, Ethiopia, the USA, Canada, Brazil, South Africa, New Zealand, the Netherlands, Kenya, and Botswana—affiliated with universities, research organizations, and international bodies—delivered lectures in person, virtually, or via hybrid formats. This approach underscored the importance of diverse expertise in guiding young scientists through the design, execution, and presentation of multidisciplinary research. In addition, CLIFOOD scholars received extensive soft skill training covering scientific presentation, scientific writing, intercultural communication, research problem-solving, and specialized statistical methods and modeling.

The design of CLIFOOD's structured education program highlights the importance of partnerships between academia, research organizations, and development practitioners, demonstrating a comprehensive approach to higher education within the context of the SDGs. This approach is evident in CLIFOOD's block seminars, which covered essential topics such as the state of the art on climate change and food security, sustainable crop production and protection, livestock systems adaptation, land–atmosphere interactions, food and nutrition security, land degradation and rehabilitation, as well as the synergies and trade-offs of climate change adaptation.

Scientific Achievements

Scholars and supervisors within CLIFOOD have produced a substantial body of scientific outputs, including peer-reviewed journal articles, oral and poster presentations, and Ph.D. theses. These contributions have enriched the knowledge base across various disciplines such as climate modeling, seasonal forecasting, sustainable farming and intensification, cropping systems, the socio-economic conditions of farmers and livestock herders, and food and nutritional security.

High-resolution climate modeling studies by Rettie et al. (2022, 2023a, b) and Kebede et al. (2024) have notably improved climate projections and seasonal rainfall predictions, providing a more reliable basis for agricultural planning and adaptation strategies in Ethiopia. Complementary analyses on climate variability and its implications for crop productivity, conducted by Habte et al. (2020, 2021), Ware et al. (2022), Senbeta et al. (2024), and Dejene et al. (2022), have deepened understanding of regional climatic impacts on agricultural systems and food security.

In sustainable farming and cropping systems, detailed research by Yimer et al. (2022, 2024a, b) has optimized maize–bean intercropping systems and fertilization practices, demonstrating practical methods to enhance productivity sustainably. Studies by Ayalew et al. (2021, 2022a, b) have highlighted cowpea's resilience under elevated UV-B radiation and water-stress conditions, promoting its integration as a viable climate-adaptive crop. Meanwhile, Gardi et al. (2021, 2022a, b) have contributed critical insights on barley yield responses to elevated CO_2 and changing climatic conditions, providing evidence to inform sustainable intensification.

Socio-economic dimensions of farming have also been significantly addressed. Research by Kayamo et al. (2023) showed how seasonal forecasts could financially benefit farmers by guiding cultivar choice, reducing economic risks associated with climatic uncertainty. Abate et al. (2024a, b) explored community-level ecosystem management, demonstrating how prescribed fire, selective thinning, and targeted grazing can effectively control woody encroachment and benefit pastoralist livelihoods.

In the livestock sector, Bayssa et al. (2021) analyzed production and adaptation characteristics of Boran cattle under changing climate conditions, complementing the comprehensive meta-analysis by Belete et al. (2024) on the nutritional value of indigenous browse species and forage legumes. Tadesse et al. (2022) further demonstrated strategies to mitigate methane emissions from ruminants without compromising productivity, an essential step toward sustainable livestock management.

Food and nutritional security have been tackled in studies by Desta et al. (2024a, b) and Ayele et al. (2022), focusing on improving nutritional outcomes using locally available, climate-resilient crops and analyzing climatic factors affecting child growth. These findings offer direct, practical insights for policy development and nutritional interventions, especially in vulnerable agro-pastoral regions.

In addition to the outputs derived from Ph.D. research, postdoctoral scholars are engaged in producing further scientific analyses aimed at evaluating the overall achievement of CLIFOOD's objectives. This collective body of work underscores the importance of integrated, climate-informed approaches to agriculture and nutrition, contributing valuable knowledge that supports sustainable development in Ethiopia and beyond.

Collectively, these diverse scientific contributions reflect CLIFOOD's holistic approach to addressing complex food security and climate resilience challenges, emphasizing context-specific, evidence-based solutions relevant to both local and global sustainability efforts.

Visibility

CLIFOOD has leveraged multiple platforms to enhance its visibility. Scholarship announcements have been disseminated through networks such as the DAAD, university consortia, embassies, and partner institutions. Furthermore, CLIFOOD has actively participated in major scientific conferences where scholars and professors have presented their work through oral presentations and posters, duly acknowledging the support of CLIFOOD, DAAD, and the Government of Germany.

Meetings of the DAAD SDG Graduate School Network have also played a key role in increasing CLIFOOD's profile. Notable gatherings include the inaugural session in Bonn (2017), a subsequent meeting in Hawassa organized by CLIFOOD (2017), and a third meeting in Vietnam (2019).

Beyond the immediate graduate school community, the CLIFOOD network encompasses a broad spectrum of external experts and institutions from academia, research, politics, business, and civil society. This extensive network supports the program's contribution to research capacity development through both South-North and South-South cooperation. The implementation of science-based block seminars has proven valuable not only for CLIFOOD scholars but also for researchers and staff at HU and UHOH.

Moreover, administrative and management capacities at HU have been strengthened through exchange programs. Engagement with a diverse range of stakeholders has also fostered new employment opportunities for young scientists, who benefit from increasingly comprehensive and practical training.

7 Sustainability and Challenges in Capacity Building

Sustainability Issues as Capacity is Built, and Demand is Created

The notable achievements of CLIFOOD have been showcased on both national and international stages, extending well beyond its block seminars and workshops. For example, CLIFOOD has been featured in forums such as RUFORUM in Cameroon (2023), at Yaoundé University in Cameroon (2023), the Water Security and Climate Change Conference in Kenya (2018), the Food Security International Conference in South Africa (2017), and multiple Tropentag conferences (2018–2024). Audience feedback has praised the unique structure of CLIFOOD and recommended its broader promotion.

A dedicated steering committee within CLIFOOD is actively exploring sustainable pathways to ensure the program's continuation beyond the project's lifespan. Their work builds on the foundational achievements already secured, including infrastructure development, curriculum design, and valuable practical experiences. Significant infrastructural investments—such as advanced laboratory equipment, a comprehensive data center with robust server facilities, and innovative eLearning materials like smart classrooms and new devices—are set to support CLIFOOD's long-term viability, even without external funding. Moreover, the meticulously developed curricula for block seminars and soft skill training are designed to remain relevant over time. Although mobility budgets present a potential challenge, CLIFOOD has created the flexibility to transition block seminars to online platforms as needed.

The steering committee is committed to ensuring that CLIFOOD continues to benefit both the participating institutions and the broader Eastern Africa region. This commitment to sustainability is further evidenced by the recent launch of a Ph.D. program in SDGs at a local Ethiopian university, funded through local sources. Such initiatives demonstrate the replicability and enduring impact of the CLIFOOD model, ensuring its continued service to Ethiopia and the wider region.

Challenges

CLIFOOD has faced a wide range of challenges, from unforeseen climatic events and global health crises to infrastructural instabilities, political unrest, economic risks, staff departures, and security concerns in Ethiopia. These issues have significantly disrupted field research and experiments conducted by young scientists, adversely affecting data collection, qualification programs, research activities, and on-site teaching.

A prime example was the COVID-19 pandemic, which severely restricted the mobility of scientific staff and disrupted qualification programs, research initiatives, and on-site teaching due to government-imposed lockouts and contact bans. As a result, several project activities were either cancelled, postponed, or transitioned to digital formats.

Administrative hurdles also emerged, particularly the challenge of reconciling DAAD financial regulations with those of HU. To address this, an international financial management training course was held at HU in November 2018. DAAD experts provided comprehensive guidance on program reporting and financial regulations, helping FSC/CLIFOOD staff and HU's finance department better understand and comply with project requirements. The workshop also facilitated discussions on communication and organizational challenges, leading to the identification of viable solutions for smoother project implementation.

The political climate in Ethiopia further complicated operations. In 2017, security concerns initially hindered researchers from accessing remote sites; however, all planned activities were successfully executed in 2018 once conditions improved. Despite subsequent stabilization in some areas, ongoing conflicts in regions such as Tigray, Amhara, and Oromia continue to pose serious security risks, affecting field research and travel, and disrupting communication and transport services—including domestic air travel.

Infrastructure instability—including power outages, limited internet access, and road damage—has complicated the implementation of qualification programs, field research, and technical capacity-building initiatives. Moreover, the slow pace of digital transformation at universities has added another layer of challenge. Limited adoption of online formats and digital teaching methods hampers the rapid dissemination of knowledge and constrains the flexibility needed to adapt to evolving technological demands. As a result, these factors not only impact the delivery of educational programs but also restrict broader access to quality education and international collaboration, further impeding innovation and progress in academic research and capacity building. CLIFOOD has responded by enhancing technical equipment, adopting blended learning formats, and adjusting schedules as needed. Although internal network disputes are relatively infrequent given the strong relationships among scientists, mediation strategies have been implemented to resolve any conflicts that arise.

In summary, these challenges—from unexpected climatic and health emergencies to persistent issues like political instability, security risks, administrative hurdles, and visa application processes—have forced CLIFOOD to adapt continuously. Through measures such as adapting activity formats and locations, enhancing technical infrastructure, and providing mediation for internal conflicts, CLIFOOD has demonstrated resilience and adaptability while maintaining its research and other activities.

Overcoming Challenges: Enhancing Quality Education and Research at HU

Despite being Ethiopia's leading agricultural institution, HU faces significant challenges in enhancing the quality of education, research, and community services. Key shortcomings include inadequate laboratory and research facilities, limited access to ICT-based teaching resources and libraries, insufficient staff numbers and qualifications, and unstable internet and database access. Furthermore, a mismatch between the number of Ph.D. candidates and available supervisors—compounded by a shortage of qualified management staff—hinders effective supervision.

SDG graduate schools like CLIFOOD are pivotal in overcoming these infrastructural and human resource challenges. Through South-South and South-North cooperation, CLIFOOD facilitates the exchange of expertise via joint teaching, supervision, and research. Moreover, the program actively promotes stakeholder collaboration by engaging a broad network of external experts from academia, politics, business, and civil society. This integrated approach represents an effective strategy for addressing HU's persistent challenges and advancing its educational and research capacities.

8 Conclusion

CLIFOOD provides a compelling model for how higher education institutions can address the pressing challenges of climate change and food insecurity in developing countries such as Ethiopia. By leveraging strong international partnerships, particularly between Hawassa University and the University of Hohenheim, CLIFOOD integrates interdisciplinary teaching, innovative research, and comprehensive capacity building under one cohesive framework. This approach has strengthened research capabilities, enhanced academic programs, and produced graduates who are well-prepared for leadership roles in academia, policy, and industry.

A key insight from CLIFOOD's experience is that climate-resilient agriculture must be grounded in both scientific rigor and social relevance. Through demand-driven research projects, scholars develop climate-smart agricultural practices adapted to local environments. At the same time, collaborative activities—ranging from block seminars and digital learning to joint field research—foster cross-cultural understanding and cultivate a broader perspective on global development challenges.

Despite obstacles such as administrative hurdles, security concerns, and infrastructural shortcomings, CLIFOOD's adaptability and resilience demonstrate the viability of international graduate schools in rapidly changing contexts. CLIFOOD's achievements, including new knowledge products, improved teaching methods, and strengthened institutional capacities, highlight the potential for sustainability beyond external funding. In turn, these outcomes illustrate that well-designed, inclusive collaborations can produce impactful, locally relevant solutions that directly contribute to the achievement of the SDGs.

References

Abate T, Abebe T, Trydte A (2024a) Do we need post-tree thinning management? Prescribed fire and goat browsing to control woody encroacher species in an Ethiopian savanna. Pastoralism Res Policy Pract 14:13039. https://doi.org/10.3389/past.2024.13039

Abate T, Abebe T, Treydte A (2024b) How much to cut? Finding an optimal thinning intensity of encroaching woody species for the herbaceous community in an East African savannah. Front Ecol Evol 12. https://doi.org/10.3389/fevo.2024.1461573

Akachi Y, Goodman D, Parker D (2009) Global climate change and child health: a review of pathways, impacts and measures to improve the evidence base. Innocenti Discussion Paper No. IDP 2009-03. UNICEF Innocenti Research Centre, Florence, 2

Araro K, Legesse SA, Meshesha DT (2020) Climate change and variability impacts on rural livelihoods and adaptation strategies in southern Ethiopia. Earth Syst Environ 4:15–26. https://doi.org/10.1007/s41748-019-00134-9

Ayalew T, Yoseph T, Cadisch G (2022a) Elevated UV-B radiation depressed biomass yield and symbiotic N_2-fixation in Bradyrhizobium inoculated cowpea varieties. Symbiosis 87:201–211. https://doi.org/10.1007/s13199-022-00868-7

Ayalew T, Yoseph T, Cadisch G (2022b) Carbon assimilation and water-use efficiency in cowpea varieties inoculated with Bradyrhizobium, measured using ^{13}C natural abundance. J Plant Interact 17(1):853–860. https://doi.org/10.1080/17429145.2022.2075943

Ayalew T, Yoseph T (2022c) Cowpea (Vigna unguiculata L. Walp.): a choice crop for sustainability during the climate change periods. J Appl Biol Biotechnol 10(3):154–162. https://doi.org/10.7324/JABB.2022.100320

Ayalew T, Lankamo M, Yoseph T (2021) Assessment of the production and importance of cowpea (Vigna unguiculata (L.) Walp.): cases from selected districts of southern Ethiopia. Afr J Food Agric Nutr Dev 21(11). https://doi.org/10.18697/ajfand.102.19630

Ayele DA, Teferra TF, Frank J, Gebremedhin S (2022a) Optimization of nutritional and functional qualities of local complementary foods of southern Ethiopia using a customized mixture design. Food Sci Nutr 10:239–252. https://doi.org/10.1002/fsn3.2663

Bayssa M, Yigrem S, Betsha S, Tolera A (2021) Production, reproduction and some adaptation characteristics of Boran cattle breed under changing climate: a systematic review and meta-analysis. PLoS ONE 16(5):e0244836. https://doi.org/10.1371/journal.pone.0244836

Belete S, Tolera A, Betsha S, Dickhöfer U (2024) Feeding values of indigenous browse species and forage legumes for the feeding of ruminants in Ethiopia: a meta-analysis. Agriculture 14:1475. https://doi.org/10.3390/agriculture14091475

Byerlee DR, De Janvry AF, Klytchnikova II, Sadoulet EM, Townsend R (2008) World development report 2008: agriculture for development. World Bank Group, Washington, D.C. Available at: http://documents.worldbank.org/curated/en/587251468175472382/World-development-report-2008-agriculture-for-development

CSA (Central Statistical Agency) (2020) Agricultural Sample Survey: Volume I – Report on Area and Production of Major Crops (Private Peasant Holdings, Meher Season), 2019/20 (2012 E.C.). Statistical Bulletin. Addis Ababa: Federal Democratic Republic of Ethiopia, CSA, 2020

CSA (Central Statistical Authority) (2021) Agricultural Sample Survey 2020/21 (2013 E.C). Report on Livestock and Livestock Characteristics. Volume II, Statistical Bulletin 589, Central Statistical Authority, Addis Ababa, 13

Dejene A, Samago TY, Dejene TA (2022) Trends in irrigated crop production in Ethiopia during the period of sustainable development goals. Preprints 2022110468. https://doi.org/10.20944/preprints202211.0468.v1

Desta DT, Tefera TF, Gebremedhin S (2024a) The effect of rainfall and temperature patterns on childhood linear growth in the tropics: a systematic review and meta-analysis. Int J Environ Res Public Health 21:1269

Desta DT, Tefera TF, Gebremedhin S (2024b) Characterization of complementary feeding practice and locally available climate-resilient crops for complementary food among agro-pastoralists of Ethiopia: a qualitative study. J Nutr Sci 13:e39

Ethiopia 2017 Voluntary National Review on SDGs (2017) Government commitments, national ownership and performance trends. Federal Democratic Republic of Ethiopia, National Planning Commission

FAO (2018) The future of food and agriculture: alternative pathways to 2050. Food and Agriculture Organization of the United Nations, Rome, p 228

Gardi MW, Memic E, Zewdu E, Graeff-Hönninger S (2022b) Simulating the effect of climate change on barley yield in Ethiopia with the DSSAT-CERES-Barley model. Agron J 114:1128–1145. https://doi.org/10.1002/agj2.21005

Gardi MW, Malik WA, Haussmann BIG (2021) Impacts of carbon dioxide enrichment on landrace and released Ethiopian barley (Hordeum vulgare L.) cultivars. Plants (Basel) 10(12):2691. https://doi.org/10.3390/plants10122691

Gardi MW, Haussmann B, Malik W, Högy P (2022a) Effects of elevated atmospheric CO_2 and its interaction with temperature and nitrogen on yield of barley (Hordeum vulgare L.): a meta-analysis. Plant Soil 475. https://doi.org/10.1007/s11104-022-05386-5

Habte A, Worku W, Gayler S, Ayalew D, Mamo G (2020) Model-based yield gap analysis and constraints of rainfed sorghum production in southwest Ethiopia. J Agric Sci 158(10):855–869. https://doi.org/10.1017/S0021859621000435

Habte A, Mamo G, Worku W, Ayalew D, Gayler S (2021) Spatial variability and temporal trends of climate change in southwest Ethiopia: association with farmers' perception and their adaptation strategies. Adv Meteorol 2021:3863530. https://doi.org/10.1155/2021/3863530

IPCC (2023a) Summary for policymakers. In: Climate change 2023: synthesis report. In: Lee H, Romero J (eds) Contribution of working groups I, II and III to the Sixth Assessment report of the intergovernmental panel on climate change. IPCC, Geneva, pp 1–34. https://doi.org/10.59327/IPCC/AR6-9789291691647.001

Kayamo SE, Troost C, Yismaw H, Berger T (2023) The financial value of seasonal forecast-based cultivar choice: assessing the evidence in the Central Rift Valley of Ethiopia. Clim Risk Manag 41:100541. https://doi.org/10.1016/j.crm.2023.100541

Kebede A, Warrach-Sagi K, Schwitalla T, Wulfmeyer V, Abebe T, Ware M (2024) Assessment of seasonal rainfall prediction in Ethiopia: evaluating a dynamic recurrent neural network to downscale ECMWF-SEAS5 rainfall. Adv Atmos Sci 41(11):2230–2244

Legese G, Gelmesa U, Jembere J et al (2023) Ethiopia national dairy development strategy 2022–2031. Ministry of Agriculture, Federal Democratic Republic of Ethiopia, Addis Ababa, Ethiopia

Oxford Policy Management and the Horn Economic and Social Policy Institute (2018) Generation El Niño: long-term impacts on children's well-being: final report. Ethiopia Country Office, UNICEF

Matewos T (2019) Climate change-induced impacts on smallholder farmers in selected districts of Sidama, southern Ethiopia. Climate 7(5). https://doi.org/10.3390/cli7050070

National Bank of Ethiopia (2020) Annual report 2018/19. Available at: https://nbebank.com/annual-report

National Planning Commission (2017) Ethiopia 2017 voluntary national review on SDGs, Addis Ababa, June 2017. Available at: https://www.et.undp.org/content/ethiopia/en/home/library/SDGs/ethiopia-2017-voluntary-national-review-on-sdgs.html

OPHI, UNDP (2024) Global multidimensional poverty index 2024: poverty amid conflict. Oxford Poverty and Human Development Initiative and United Nations Development Programme, Oxford

Philip S, Kew S, Jan van Oldenborgh G, Otto F, O'Keefe S, Haustein K, King A, Zegeye A, Eshetu Z, Hailemariam K, Singh R, Jjemba E, Funk C, Cullen H (2018) Attribution analysis of the Ethiopian drought of 2015. J Clim 31(6):2465–2486. Available at: https://journals.ametsoc.org/view/journals/clim/31/6/jcli-d-17-0274.1.xml. Accessed 7 Apr 2025

Pritchard B, Ortiz R, Shekar M (eds) (2016) Routledge handbook of food and nutrition security. Routledge, Oxfordshire, UK, New York, USA, 524pp

Rettie FM, Gayler S, Weber TK, Tesfaye K, Streck T (2022) Climate change impact on wheat and maize growth in Ethiopia: a multi-model uncertainty analysis. PLoS ONE 17(1):e0262951

Rettie FM, Gayler S, Weber TK, Tesfaye K, Streck T (2023a) Comprehensive assessment of climate extremes in high-resolution CMIP6 projections for Ethiopia. Front Environ Sci 11:462

Rettie FM, Gayler S, Weber TK, Tesfaye K, Streck T (2023b) High-resolution CMIP6 climate projections for Ethiopia using the gridded statistical downscaling method. Sci Data 10(1):442

UN SDG (2015) United nations sustainable development goals. Available at: https://www.un.org/sustainabledevelopment

Senbeta AF, Worku W, Gayler S (2024) Spatiotemporal climate variability and food security implications in the Central Ethiopia region. Sci Afr 26:e02390. https://doi.org/10.1016/j.sciaf.2024.e02390

Tadesse A, Melesse A, Rodehutscord M (2022) Partial substitution of concentrate mix with dried Leucaena leucocephala leaf reduced in vitro methane production in rams without affecting the nutrient intake and performance traits. Trop Subtropical Agroecosyst 25(2). https://doi.org/10.56369/tsaes.4060

U.S. Department of Commerce (2023) Ethiopia—agricultural sectors. International Trade Administration. Available at: https://www.trade.gov/country-commercial-guides/ethiopia-agricultural-sectors. Accessed 19 Mar 2025

UNICEF (2007) Climate change and children: policy. UNICEF Innocenti Research Centre, New York, 6p. Available at: http://www.unicef.org/publications/files/Climate_Change_and_Children.pdf

USAID (2016) Climate change risk profile Ethiopia fact sheet. Available at: https://www.climatelinks.org/sites/default/files/asset/document/2016%20CRM%20Factsheet%20-%20Ethiopia_use%20this.pdf

Ware M, Mori P, Warrach-Sagi K, Jury M, Schwitalla T, Beyene KH, Wulfmeyer V (2022) Climate regionalization using objective multivariate clustering methods and characterization of climatic regions in Ethiopia. Meteorol Z 31(6):431–453. https://doi.org/10.1127/metz/2022/1093

WHO (2022) Greater horn of Africa drought and food insecurity grade 3 emergency: Djibouti, Ethiopia, Kenya, Somalia, South Sudan, Sudan, Uganda. Situation report. Available at: https://cdn.who.int/media/docs/default-source/documents/emergencies/who_ghoa_sitrep_2022-08-08.pdf?sfvrsn=dbdfc8b0_3&download=true

World Bank (2010) Economics of adaptation to climate change, Ethiopia. World Bank, Washington, D.C.. Available at: https://openknowledge.worldbank.org/handle/10986/12504

World Bank (2020a) Rural population (% of total population). Available at: https://data.worldbank.org/indicator/SP.RUR.TOTL.ZS?locations=ET

World Bank (2020b) World development indicators. Available at: https://databank.worldbank.org/source/world-development-indicators

World Bank (2020c) The locust crisis: the world bank's response. Available at: https://www.worldbank.org/en/news/factsheet/2020/04/27/the-locust-crisis-the-world-banks-response

Yimer T, Abera G, Beyene S, Rasche F (2022) Optimizing maize–bean cropping systems for sustainable intensification in southern Ethiopia. Agron J 114:3283–3296. https://doi.org/10.1002/agj2.21143

Yimer T, Abera G, Beyene S, Ravensbergen APP, Ukato A, Rasche F (2024b) Optimizing fertilization schemes to narrow the maize yield gap in smallholder farming systems in southern Ethiopia. Heliyon 10:e33926

Yimer T, Abera G, Beyene S, Bono B, Rasche F (2024a) Combined inoculation of arbuscular mycorrhiza fungi with Meso-rhizobium improves nutrient uptake, growth performance, and moisture stress tolerance of chickpea (Cicer arietinum L.). Agroecosyst Geosci Environ. https://doi.org/10.1002/agg2.20562

MIX
Papier aus verantwortungsvollen Quellen
Paper from responsible sources
FSC® C105338
FSC
www.fsc.org

If you have any concerns about our products,
you can contact us on
ProductSafety@springernature.com

In case Publisher is established outside the EU,
the EU authorized representative is:
Springer Nature Customer Service Center GmbH
Europaplatz 3, 69115 Heidelberg, Germany

Printed by Libri Plureos GmbH
in Hamburg, Germany